STUDIES IN HISTORY, ECONOMICS AND
PUBLIC LAW

Edited by the
FACULTY OF POLITICAL SCIENCE
OF COLUMBIA UNIVERSITY

NUMBER 483

GIOVANNI MARLIANI AND LATE
MEDIEVAL PHYSICS

BY

MARSHALL CLAGETT

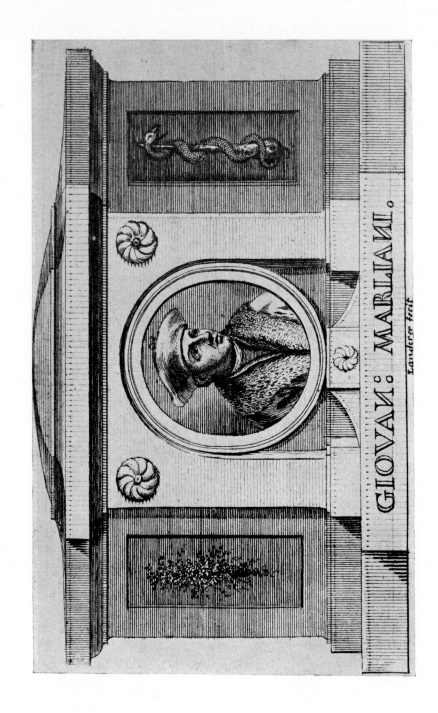

GIOVAN: MARLIANI.

Landerer fecit

GIOVANNI MARLIANI
AND
LATE MEDIEVAL PHYSICS

BY

MARSHALL CLAGETT, Ph. D.

NEW YORK

COLUMBIA UNIVERSITY PRESS

LONDON: P. S. KING & SON, LTD.

1941

To

C. M. C.

AND

B. C.

PREFACE

THIS study has as its purpose the investigation of certain physical concepts of the fourteenth and fifteenth centuries by means of the problem method. This method, so universally used in the science of today and so rarely employed in studies of late medieval thought, is necessary to provide the material from which generalizations can be drawn. Some of the material presented here will seem of little significance in the light of modern science. It is only too obvious, however, that we can never have a faithful representation of the natural philosophy of any period without presenting its " error " conjointly with its " truth." Although Giovanni Marliani is perhaps less important from the standpoint of the proportion of " truth " to " error " than some of his fourteenth century predecessors (e. g. Bradwardine, Swineshead, Dumbleton, Buridan, and Oresme), he was considered by his contemporaries as an outstanding figure in the academic world. Consequently, a study of his ideas is of some value in appreciating the general outlook of the Italian university in the middle part of the fifteenth century.

A preface should be limited largely to giving credit where it is due. I owe not an inconsiderable debt to works that have already been published, particularly to the third series of Pierre Duhem's *Études sur Léonard de Vinci*, Paris, 1913. This volume is a splendid attempt to sift from the writings of fourteenth and fifteenth century philosophers their scientific content. If at times it suffers from an over-enthusiastic evaluation of the importance of the ideas uncovered or from an occasional technical error, its general value and usefulness is not in the least impaired.

Something likewise must be said of those men who have read my study and suppressed its most patent errors. Among them are Professors Austin P. Evans, Robert von Nardroff, Dino Bigongiari, and Dr. Carl Boyer. To Professor Frederick Barry and Dr. Ernest A. Moody I owe special thanks, to the first for reading the work in typescript, to the latter for his help in interpreting the fourteenth century schoolmen. I am in greatest debt of all to Professor Lynn Thorndike. He has more than once made it possible for me to continue my work at Columbia. The training I have received as his research assistant and in his seminar, in addition to his constant personal aid, has made possible the completion of this work.

Final thanks are due to Miss Evelyn Ferry, Mrs. Manning Clagett, Mr. Francis S. Benjamin, Jr. and Mr. John Mason for their part in typing the manuscript and reading proof.

TABLE OF CONTENTS

                                                                      PAGE
GIOVANNI MARLIANI . . . .   *Frontispiece*

CHAPTER I
Life and Works   11

CHAPTER II
Reaction: Early Speculations on Heat   34

CHAPTER III
The Reduction of Hot Water   59

CHAPTER IV
Body Heat and Antiperistasis.   79

CHAPTER V
The Uniform Equivalence of Uniformly Accelerated Motion   101

CHAPTER VI
The Peripatetic Law   125

CHAPTER VII
Vulgar Fractions   145

CONCLUSIONS   168

APPENDIX   171

INDEX   177

CHAPTER I
LIFE AND WORKS

GIOVANNI MARLIANI was born of a patrician family in
Milan sometime during the first quarter of the fifteenth century when that city was under the control of the Visconti. The
Marliani family had been an important one in the history of
Milan from the early days of the commune when Alberto
Marliani aided in preparing the *Consuetudines* of 1226 [1] to
the time of Giovanni. But it was in the fifteenth and sixteenth
centuries that the family produced its most distinguished members, many of whom were closely related to Giovanni.

Little is known of Giovanni's father, except that he served
at least one of the Visconti, perhaps the great Gian-Galeazzo,
first duke of Milan, as *rationator* of the ducal chamber,[2] a
position which later passed to his son Daniele, brother of
Giovanni.[3] Daniele's career has been obscured by the brilliant
activities of his son, Luigi, who, like Giovanni, studied medicine
at Pavia and was elected to the College of Physicians of Milan.
However, unlike his uncle, he turned not to teaching, but instead became counselor and physician to some of the most
important political figures of his time: Lodovico Sforza il Moro
and Massimiliano, dukes of Milan; Maximilian and Charles V,
German emperors; and King Philip I of Spain. As a final
reward he was made bishop of Tuy.[4]

1 Filippo Argellati, *Bibliotheca Scriptorum Mediolanensium*, Vol. II, part 1,
Milan, 1475, cs. 860-861.

2 *Ibid.*, II, 1, 866.

3 *Ibid.*, II, 1, 861. That Daniele was Giovanni's brother follows from a
statement made by Zanino Volta to the effect that Luigi (Aloysius) Marliani
was the son of Giovanni's brother. Luigi's father is called Daniele by Argellati.
See Volta, " Del Collegio Universitario Marliani in Pavia," *Archivio storico
Lombardo*, Second series, Vol. IX (1892), p. 595.

4 Paolo Morigia, *La Nobilità di Milano*, Milan, 1595, p. 147. Luigi was
eulogized along with a Gianfrancesco Marliani and others by a plaque erected

Of almost equal fame was Raimondo Marliani, a cousin of
Giovanni, who as a jurist taught at Louvain and Dôle. He
was a canon of Liège and by his will in 1475 established the
Collegio Marliani at the University of Pavia.[5]

Giovanni had three sons, Girolamo, Paolo, and Pietro
Antonio.[6] Like their father they all taught at one time or another at the University of Pavia. Paolo was nominated in 1483
to read *ad libitum* mathematics, philosophy, or logic. The chair
which he occupied was one that had been established for the
family. The stipend and duties were to be shared with his

in the Chiesa Santa Maria della Pace in Milan. It reads (Argellati, *op. cit.*,
II, 1, 871-872) :

    IOH. FRANCISCO MARLIANO . . . .
      ET ALOYSIO PHILIPPI I REG. HISP.
    MAXIMILIANI ET CAROLI V CAESARUM
    MAXIMILIANI ET LUDOVICI DUCUM MEDIOLANI
      A CONSILIIS SIMUL ET ARCHIATRO
    ORATORI MATHEMATICO ATQUE PHILOSOPHO
    EPISCOPO TUDENENSI CARDINALI DES.

5 Morigia, *op. cit.*, p. 147; Argellati, *op. cit.*, II, I, 875-876; Volta, *op. cit.*,
passim.

6 It is rather strange that B. Corte (*Notizie istoriche intorno a'medici
scrittori Milanesi*, Milan, 1718, p. 34), and after him Argellati (*op. cit.*, II, 1,
866) say that Giovanni had two sons, and yet they print a plaque erected by
his grandson to Giovanni and his three sons. This plaque was placed in the
Chiesa Santa Maria delle Grazie and reads (Argellati, *op. cit.*, II, 1, 867-868) :

      IOANNI MARLIANO AVO
   CIVI MEDIOL. ET PHILOS. NOBILISSIMO
      ET PAULO PATRI
      USU PRUDENTIAQUE
   PRINCIPIBUS CIVITATIS ACCEPTISSIMO
  ET HIERONYMO ET PETRO ANTONIO PATRUIS
   OMNI LITTERARUM ELEGANTIA PERPOLITIS
    PETRUS ANTONIUS MARLIANUS
  IURISCONSULTUS ET SENATOR MEDIOLANI
   MAIORIBUS SUIS BENEMERENTIBUS
     POSUIT MDLXII

Confirmation that all three were Marliani's sons appears in the roll of professors at Pavia for the year 1486-7 (*Memorie e Documenti per la Storia
dell' Università di Pavia*, part 1, Pavia, 1878, p. 114) : " Paulus, Hieronymus,
et Petrus Antonius filii quondam Magistri Johannis de Marliano."

brothers.[7] Although it is not specifically stated which of the three subjects Paolo lectured on in 1483, according to Parodius two years later he was teaching moral philosophy.[8]

His brothers, on the other hand, appear in the University records for the first time in 1486-1487, apparently teaching *ad libitum* under the arrangements previously made for the family.[9] Pietro Antonio is listed again in the rolls of 1488 and 1499 as reading the morning course in natural philosophy; while Girolamo is found once more in a University document of 1507.[10] Both of the latter brothers are spoken of as physicians who received special privileges and praise from Lodovico il Moro and Francesco II Sforza.[11]

It appears to have been Paolo's descendants who carried on the academic and political traditions of the family through the sixteenth century. One son, Pietro Antonio, was a jurist known for his command of languages and as a counselor to Philip II of Spain. Another, Girolamo, was a military tribune. And a grandson by Pietro, Pietro Camillo Marliani, was Count of Busti, jurist, senator of Milan, and president of the *Questura ordinaria*.[12]

Although the broad outline of the activities of this fertile Milanese family can be readily traced, specific details, such as birth dates, early childhood training and education, etc., are almost universally wanting. Such is the case with Giovanni. That he studied arts and medicine at the University of Pavia, however, can not be doubted. An edict promulgated by Galeazzo II Visconti in 1361, and later confirmed by Filippo Maria Visconti when he undertook in 1412 the revitalization of the

---

7 *Memorie e Documenti*, I, p. 149.

8 Jacobus Parodius, *Elenchus privilegiorum et actuum publici Ticinensis studii*, 1753, p. 139.

9 *Memorie*, I, p. 149; Parodius, *op. cit.*, p. 139.

10 *Memorie*, I, p. 149.

11 Argellati, *op. cit.*, II, 1, 866.

12 Morigia, *op. cit.*, pp. 147-148; Volta, *op. cit.*, p. 595; Argellati, *op. cit.*, II, 1, 873-874.

university, made it necessary for all students in the duchy to pursue their studies at Pavia.[13] This, of course, included citizens of Milan.

In 1440 Marliani concluded his studies and was elected to the College of Physicians of Milan.[14] This would indicate that he had already taken his doctorate. According to Brambilla, he received the degree at Pavia in August, 1440,[15] or in other words at the same time as his election to the College of Physicians. However, Giovanni's name is not listed among the *licentiati* or *doctorati* given in the *Codice Diplomatico dell'Università di Pavia* for the year 1440,[16] or as a matter of fact for any year. Be that as it may, two years later in 1442 he is mentioned in a list of Doctors of Arts appearing as examiners of a candidate for the doctorate; while in 1443 at a similar examination he is listed among several Doctors of Arts and Medicine.[17]

Marliani began to teach at Pavia in 1441, at which time he received a salary of forty florins for the " extraordinary " lectures in natural philosophy (i.e. physics).[18] These " extraordinary " lectures were probably given in the afternoon, ac-

13 *Memorie*, II, pp. 3-4, 8-9.

14 See the fragment of the *matricula* of the Milanese College of Physicians published by Johannes de Sitonis in the supplement to B. Corte, *op. cit.*, p. 287. This college was apparently a kind of guild of physicians, and did not necessitate Marliani's residence in Milan throughout the year. The membership of the college no doubt included chiefly those who had taken their degrees at Pavia.

15 Joh. Alexander Brambilla, *Storia delle scoperte fisico-medico-anatomico-chirurgiche fatte dagli uomini illustri italiani*, Vol. I, Milan, 1780, p. 144.

16 *Codice Diplomatico*, Vol. II, part 1, Pavia, 1913, p. 410.

17 *Ibid.*, Vol. II, part 2, Pavia, 1915, pp. 447, 472.

18 *Ibid.*, II, 1, p. 433. The index of the *Codice* confuses Giovanni with his namesake, who also taught philosophy at Pavia. However, this other Giovanni was an Augustinian monk, who appears in the list of professors for the year 1418. Neither Parodius (*op. cit.*, p. 139) nor the editor of the *Memorie* (I, pp. 114, 155) makes this mistake. Marliani conclusively confirms 1441-1442 as his initial year when he writes in 1472 that he has taught at Pavia for thirty years (*Questio de caliditate corporum humanorum*, Milan, 1474, f. 1r, col. 1).

cording to the custom at Bologna. As their name implies, they were additional readings and were most often put in the hands of lesser known or newly confirmed licentiates. The young physician seems to have given satisfaction during this first year, for on January 16, 1442, Filippo Maria Visconti, who as Duke of Milan and Count of Pavia exercised strict control over the affairs of the University, granted to Marliani an additional twenty florins.[19] That the Duke was further pleased by Marliani's lectures is evident from the fact that in the academic year of 1443-1444 Giovanni was given the " ordinary " (morning) lectures in philosophy in addition to some readings on astrology (no doubt on feast days) at an increased salary totaling one hundred florins.[20] Marliani's steady rise in salary continued until two years later (1446-1447) when he received a stipend of two hundred florins for the same courses.[21]

At the close of the year 1446-1447 a political event took place which altered the course of Marliani's life somewhat. This was the death of Filippo Maria Visconti. Since Filippo was the last male descendant of Gian-Galeazzo Visconti, there was no evident successor to the duchy. The people of Milan saw in this circumstance the opportunity to recapture their liberty. They demolished the Castello di Porta, symbol of Visconti control, and at the same time set up the so-called *Republica Ambrosiana,* which was to be administered chiefly by a *corpo di ventiquattro amministratori.* The membership of this body was later reduced to twelve and took the name *I Capitani e Defensori della Libertà.* Meanwhile other parts of the duchy fell into the hands of Francesco Sforza, son-in-law of the late duke, among them the city of Pavia with its university. And

19 *Ibid.,* II, 2, p. 443. "Addimus preterea magistro Johanni de Marliano viginti florenos in anno, ita ut et ipse, computatis florenis quadraginta sibi dessignatis pro lectura ad quam deputatus est, habere veniat annuatim in totum sexaginta florenos."

20 *Ibid.,* II, p. 470 "Ad lecturam Philosophie ordinariam: . . . M. Johannes de Marliano, flor. centum . . . Ad lecturam Astrologie: . . . M. Johannes de Marliano pro salario supra."

21 *Ibid.,* II, 2, p. 496.

although Sforza was officially appointed Captain-General of the Republic, one of the principal tasks of the *Capitani* was to prevent Milan likewise from falling to the ambitious Sforza. Hence relations between Pavia and Milan were strained. All citizens, including members of the university, were called home to Milan.[22]

In the month of October, 1447, in line with their policy of maintaining the independence of Milan culturally as well as politically, the *Capitani* appointed a committee to set up a *studium generale* which was to include faculties of every kind.[23] This committee, so we are told by its own members, drew up a roll of professors and courses. We may safely assume that Marliani was listed in this original roll, for when in April, 1448, a list of twenty-two instructors for the following year was compiled, Marliani's name appeared along with five other Milanese citizens who had taught at Pavia.[24]

Marliani's recall from Pavia acted to his advantage so far as his courses were concerned. At the new university he had

22 This can be seen from a document dated November 21, 1447, which granted to a Giovanni Vimercati the privilege of remaining at Pavia in spite of the general recall. *Codice Diplomatico*, II, 2, p. 515, " Capitani et defensores libertatis . . . Communitatis Mediolani. Quamquam mandaverimus per publicas proclamationes unumquemque subditum huius inclite civitatis, existentem extra hanc civitatem, ad eam redire infra certa tempora in ipsis proclamationibus limitata, tamen ut requisitionibus concivis nostri carissimi domini Iohannis de Vicomercato Iuris utriusque doctoris, actu presentialiter legentis in civitate Papie, annuamus, contenti sumus et licentiam sibi concedimus quod ipse tute, libere, et impune in dicta civitate Papie morari et stare possit..."

23 *Ibid.*, II, 2, p. 527 " Usque in mense octubri proxime preterito, ordinatum fuit per Illustres Capitaneos...debere per nos infrascriptos erigi generale Studium in quacumque facultate..."

24 This roll of 1448 is attached to a document assigned by the original founding committee and dated April 27, 1448 (*Codico Diplomatico, loc. cit.* in note 23). The members of the committee discuss the foundation, remarking that their instructions have not been carried out precisely as they wished. Some of the professors have not lectured a sufficient period of time, and hence one-fourth must be deduced from the salaries of the offending professors. It seems likely that Marliani escaped any reduction in salary, for he is listed in the roll as receiving two hundred florins, which was the same stipend he had had at Pavia.

the opportunity to give the "ordinary" lectures in medicine instead of natural philosophy. As before at Pavia he likewise read astrology on feast days.[25]

Up to this time Giovanni's whole interest had been concentrated on natural philosophy or physics. It is to the pre-Milan period at Pavia (1441-1447) that Marliani refers in his *Questio de proportione motuum in velocitate* when he speaks of having read publicly the treatises of Bradwardine and Albert of Saxony on the proportion of motions.[26] Although Marliani taught medicine from 1447 until almost the end of his life, never did that interest in physics developed in this early period at Pavia flag. In fact, all of the physician's physical works were written during the later period.

In addition to Marliani's duties as professor of medicine at the University of Milan, he likewise assumed the position of *decurio patrie*.[27] However, both this political venture and his academic career in his native city were shortlived, for in 1450 the Ambrosian Republic collapsed before Francesco Sforza, and the University of Milan, scarcely three years old, was closed at the same time.

Not long after the fall of the republic Marliani returned to the University of Pavia, perhaps in the same year.[28] At any rate, his name is found in a Pavian roll of 1453 *ad lectorem physice et astrologie* and again in 1455 as morning lecturer in medicine and astrology.[29] The Milanese physician was not only none the worse for his brief stay at the University of

25 *Ibid.*, II, 2, p. 528.

26 See Chapter VI *infra*, note 30.

27 Corte, *op. cit.*, p. 274.

28 There is a document dated November 14, 1450 and cited by Parodius (*op. cit.*, p. 35) which provided for an increase in the number of professors at Pavia. This document has not been published, and hence I am unable to determine whether Marliani's name appears in it. It would seem likely that Marliani returned to Pavia at this time, for the next document providing for an increase in the staff is dated 1454 and we have evidence that Marliani was at the university in 1453 (see note 29 *infra*).

29 *Memorie*, I, p. 114.

Milan, but actually was better off. Instead of returning to his old lectures on philosophy, he continued those on medicine which he had inaugurated at Milan. Medicine was the future for one teaching philosophy.

The next reference to Marliani at Pavia is in 1457 when one of his colleagues, Ferrari da Grado, complained to the Duke that he received only 14 grossi for his part in medical examinations, while Marliani and others received one florin.[30] To this reference indicating Marliani's participation in the academic routine can be added a document in which Marliani is listed among those who attended in 1459 a convocation of the College of Artists and Doctors to amend the statutes for electing a prior.[31]

It is not until 1463 that we have any indication of Marliani's salary since his return to Pavia. In that year he was given five hundred florins for the morning lectures on medicine and his customary readings on astrology.[32] We may assume that he received this same salary for some time, since an expense account for 1467 indicates a similar stipend.[33] During these years when Marliani received just about the top salary, he seems to have been in the special favor of Francesco Sforza. A letter dated one month before the latter's death in March, 1466, grants the physician an additional 150 florins. The grant was made secretly so that the other professors would not be envious.[34]

30 H. Ferrari, *Une chaire de médecine au XVe siècle* (Ferrari da Grado at Pavia), Paris, 1899, p. 39.

31 *Statuti e Ordinamenti della Università di Pavia*, Pavia, 1925, p. 134.

32 An expense account for 1463 is published by M. Formentini, *Il Ducato di Milano*, Milan, 1887, p. 633. Unfortunately the period after 1450 at Pavia is not as well covered by published materials as the preceding one. The *Codico Diplomatico* extends only to 1450.

33 H. Ferrari, *op. cit.*, p. 47.

34 This letter is noted in Parodius, *op. cit.*, p. 39. It has been printed in part by Mariano Mariani in his *Vita Universitaria Pavese nel secolo XV,* Pavia, 1899, p. 119.

Perhaps Marliani's most significant academic success was achieved three years after the Duke's death when in 1469 he assumed the chair of theoretical medicine, which among medical courses was the most coveted. He is listed in the record as still teaching this course in 1483, the year of his death.[35] However, as we shall see, he must have spent considerable time away from the university in the intervening period.

Although the evidence of his activities other than the composition of certain works is scarce for the decades 1460-1470 and 1470-1480, we catch brief glimpses of friendships with Francesco Filelfo, the noted humanist, Bernardo Torni, the Florentine physician, and a number of Giovanni's own colleagues at the University of Pavia.

Two letters from Francesco are all that remain of his correspondence with Marliani. In the first one, dated 1463,[36] Francesco asks Giovanni to permit him to see a work of (an unknown) Johannes Euglyphus, which their mutual friend Ambrogio Griffo had told him that Giovanni possessed. The humanist promises to return it quickly. The second letter, written in 1466,[37] reminds Marliani that when he last left Milan for Pavia, he had promised to send Francesco some material from the *Problemata* of Alexander of Aphrodisias. Filelfo further requests that Marliani indicate all that Albertus Magnus had written on the nature of the gods. In order that Marliani will not think this a strange request, Francesco explains that, as Marliani knows, the works of Albertus are conserved at Cologne, and it has come to his attention that there are a number of German students at Pavia. Therefore, it would be comparatively easy for Marliani to secure the desired titles from these students.

Even slighter is the information which connects Bernardo Torni with Marliani. The Florentine declares in his *Annotata* to William Heytesbury's *Tria predicamenta de motu* that

35 *Memorie*, I, p. 114.

36 F. Filelfo, *Epistolarum familiarum libri XXXVII*, Venetiis, 1502, f. 152v.

37 *Ibid.*, f. 185v.

Marliani has written explaining his opinion on a case involving local motion.[38]

From his own university Marliani cites Ambrogio Griffo and Lazaro Tealdo as those who have been of constant aid to him,[39] while he dedicates his *Questio de proportione* to the important Pavian doctor and ducal physician, Benedetto Reguardati, with whom he no doubt often came in contact. Finally we may mention from the University of Pavia Ferrari da Grado, the eminent physician, who in 1468 together with Marliani was commissioned to preserve the public health in Pavia.[40] Although they were not, strictly speaking, friends of the Milanese physician, Gaetan of Tiene, Giovanni Arcolani, and Philip Adiuta may be said to have been acquaintances of his in the sense that they engaged with him in written dispute.[41]

Some light is cast on the last decade of Marliani's life by a decree of the young Duke Gian-Galeazzo Maria Sforza dated December 23, 1482.[42] The decree first remarks on Giovanni's long career at the University of Pavia and calls him " our dearest physician," " mathematician and philosopher supreme," and " noted and distinguished Doctor of Arts and Medicine." It goes on to state that because of the incredible rewards that accrue to those who are students under Marliani, his name has become famous not only in all of Italy, but in other countries as well. His fame is such that students who wish to be instructed in medicine, philosophy, or mathematics would think of choosing almost no other instructor. In fact, Giovanni is

38 The *Annotata* printed with Heytesbury's *Tractatus de sensu composito et diviso* (et al.), Venetiis, 1494, f. 74r.

39 See Chapter IV *infra*, note 9.

40 H. Ferrari, *op. cit.*, pp. 57, 316.

41 See Chapters II, III, and VI *infra*.

42 This decree is published almost in full (except for the actual gifts and privileges extended to Marliani and his heirs) by B. Corte, *Notizie istoriche*, pp. 31-33. I suspect that at least part of the grant made by this decree was the creation of a chair at the University of Pavia for Marliani's family (see note 7 and text *supra*).

described as another Aristotle in philosophy, another Hippocrates in medicine, and another Ptolemy in astronomy.

Following this laudatory introduction, we are told that Marliani was called away from Pavia to care for the health of Gian's father, Galeazzo Maria Sforza. The probable date of this summons is 1472, or shortly before, since in the introduction to his *Questio de caliditate,* written in that year, Marliani speaks of having left Pavia at the order of Galeazzo Maria (see note 18 *supra*). It is quite likely that this retirement from Pavia was more or less permanent, although, as we have seen, the physician was mentioned as teaching theoretical medicine as late as 1483 (see note 35 *supra*). At any rate, Gian's decree further remarks that Marliani was truly patriotic in remaining in his native city and resisting the pressure of those in Venice, Bologna, Siena, Perugia, and other places, who were willing to pay him handsomely to quit Milan in their favour. After the murder of Galeazzo Maria, Marliani continued in the service of his six-year-old son, Gian-Galeazzo, as his personal physician. The decree declares that Marliani could not have taken better care of Gian had the latter been the physician's own son. In return for this service Giovanni is granted certain gifts and privileges.

Almost a year later on September 26, 1483 this decree was renewed and confirmed by the young duke. In the confirmation Gian remarks that at that moment Marliani lies critically ill. He hopes for a speedy recovery.[43] However, Giovanni failed to regain his health and died sometime later in the year.[44] He was buried in Santa Maria delle Grazie in Milan.[45]

The famous physician was remembered in his native city for his own achievements and those of his progeny, and through-

---

43 This confirmation has also been published by Corte, *op. cit.*, pp. 33-34.

44 Donato Bosso in the *Chronica Bossiana* (Milan, 1492) under the year 1483 cites the death of Marliani as occurring on September 21. If the confirming decree noted above is dated correctly, we know that Marliani, although seriously ill, was still alive on September 26. *Cf.* Girolamo Tiraboschi, *Storia della Letteratura Italiana*, Vol. IV, Rome, 1783, pp. 403-404.

45 See note 6 *supra* for the plaque erected by his grandson.

out Italy for his students (such as George Valla) and for his writings, which were read by such eminent successors as Leonardo da Vinci, Alessandro Achillini, and Galileo.

II. Works

It is the purpose of this section to list the manuscripts and editions of all of Marliani's works. In addition, I have included here discussions of the dates of the various compositions. Cross references should connect this material sufficiently with the topical studies of the succeeding chapters.

A. TRACTATUS DE REACTIONE

There are three manuscripts and one printed edition of this work. Two of the manuscripts are in codices at the Biblioteca Nazionale di San Marco in Venice (cod. VI, 105, ff. 23-41 and VI, 30, ff. 21-43), while the third is part of the Marliani codex in the Aldini collection of the R. Biblioteca Universitaria di Pavia (cod. 314, ff. 90-108). The printed copy is in the second volume of the incunabular edition of Marliani's works (see section M, *Disputatio* etc., ff. 25-51). San Marco VI, 105 has been used in the preparation of Chapter II of this study.

The *Tractatus de reactione* was composed in 1448 at Milan. The *explicit* in each manuscript is the same as regards the date of composition: " Explicit tractatus in materia de reactione, compositus per (famosissimum et astronomorum verissimum dominum magistrum) Joannem de Marliano, Mediolanensem, Artium et Medicine doctorem (et philosophie sacrae monarcham) in civitate Mediolani, anno Domini nostri Ihesu Christi 1448, dum febre quartana moleste affligeretur." The words in parentheses are those added by the copyist in San Marco VI, 30, f. 43v. In his *Bibliotheca Manuscripta ad S. Marci Venetiarum* (Vol. IV, p. 164) Valentinelli has read the date of this last mentioned manuscript as 1444, but the Bibliotecario Direttore of the library, Dr. Zorzanello, has kindly informed me by letter that Valentinelli has misread the date, that actually it is 1448. Finally, Marliani has confirmed the date as 1448.[46]

The only modern writer to discuss the *Tractatus de reactione* is Pierre Duhem in his *Études sur Léonard de Vinci,* III, Paris, 1913, pp. 497-498. His discussion is so hopelessly erroneous that it indicates he had not read the work; otherwise it is inconceivable that a historian of his acumen could have made such patent errors. Duhem commences with a discussion of a treatise on reaction by Gaetan of Tiene, which he cites in an edition of 1525 of Gaetan's works. This treatise, Duhem notices, opens with an introductory statement to the effect that it does not pretend to be a complete discussion of the problem of reaction, but merely a study of (1) a theory which had appeared in a recently composed treatise on the subject and of (2) his own theory as already advanced in his commentaries on the third book of Aristotle's *Physics.*

In Gaetan's statement we can discover no indication of the name of the author of the " recently composed " treatise to which he refers. Immediately, however, Duhem jumps to the false conclusion that this unknown author is the Calculator (Richard Swineshead, whom Duhem erroneously believed to be Richard de Ghlymi Eshedi) [47] and that Gaetan's discussion of reaction is an attack on the views of the Calculator. After committing this fundamental error, the Frenchman notices from the incunabular edition of Marliani's works (probably from a description of it in Hain) that there is a *De reactione* by Marliani, an answering *De reactione* by Gaetan, and a final response by Marliani. So, apparently without reading this edition, Duhem assumes that Marliani's treatise is a defense against the " original " attack made by Gaetan in the *Tractatus de reactione,* which was cited in the edition of 1525 already mentioned, and that the response by Gaetan in the Marliani incunabulum is a second essay on reaction by the Paduan doctor, answered in turn by a second treatise by Marliani.

46 *Questio de caliditate corporum humanorum,* Milan, 1474, f. 9v, c. 2.

47 See the appendix *infra* for a discussion of Swineshead.

These assumptions are negated by the following considerations: (1) The so-called "first" treatise *De reactione* of Gaetan included in the edition of 1525 proves upon examination to be identical with the "second" one in the Marliani incunabulum. (2) Marliani's first work on reaction is not a defense of the Calculator against an attack by Gaetan, as Duhem alleges, but rather a complete examination of the question, which in part, at least, is in opposition to the Calculator.[48] The only mention made by Marliani of Gaetan is in the third part where he briefly refutes the latter's opinions as contained in his commentaries on the *Physics* rather than in any separate treatise attacking the Calculator.[49] (3) Since, as we have seen, Gaetan's "two" treatises turn out to be only one, it remains to remark that this one work was written after Marliani's first essay. It is a point by point attack on Marliani's theory of reaction and a detailed defense of his own theory against the criticisms of Marliani.[50] The "recently composed" treatise referred to by Gaetan is of course that of Marliani.

B. IN DEFENSIONEM TRACTATUS DE REACTIONE

The manuscript and printed editions of Marliani's defense of his first study of reaction are the same as those of the earlier work: San Marco codices VI, 105, ff. 1-7 and VI, 30, ff. 158-168; R. Bibl. Univ. di Pavia, Aldini 314, f. 114 ff.; and *Disputatio* etc. (See section M), f. 59v ff. San Marco VI, 105 will be cited in Chapter II *infra*.

The date of composition of this work is somewhat confused, since each edition bears a different date, and a reference by Marliani to the year in which it was composed agrees with none of those in any of the editions. San Marco VI, 105 is dateless, while San Marco VI, 30, has 1444.[51] The Pavia

---

48 See Chapter II *infra*, note 57 and text.

49 Cee Chapter II *infra*, notes 50, 51, *passim*.

50 See Chapter II *infra*, notes 55, 59, 61, etc.

51 Dr. Zorzanello has confirmed by letter Valentinelli's reading of 1444.

codex gives 1454 [52] and the printed edition 1467. Marliani, however, declares that he wrote the *Defense* in 1456.[53]

We may conclude that 1444 and 1467 are out of the question, the former because this treatise is a defense of one written in 1448 and the latter because it is cited in Marliani's *Disputatio cum Joanne de Arculis*,[54] which was written in 1461 (See section F *infra*). Consequently the *Defense* must have been written in either 1456, as Giovanni tells us, or, if by chance his memory was faulty, in 1454, as the Pavia codex reveals.

C. TRACTATUS PHYSICI (LIBER CONCLUSIONUM DIVERSARUM?)

The *Tractatus physici* appear exclusively in the Pavian codex mentioned in the preceding sections (Aldini 314, ff. 1-83r).[55] A comparison of the chapters of this work with those of the *Liber calculationum* of Richard of Swineshead makes it evident that Marliani has here composed a commentary on the treatise of the Oxford schoolman. Each division of the *Tractatus physici* can be found likewise in the *Liber calculationum*.[56] The table of contents of Marliani's work as printed

52 This codex (Aldini 314) bears the stamp of the Biblioteca dei Minori Conventuali di S. Francesco of Milan. Argellati (*op. cit.*, II, 1, 867) mentions reading a copy of the *Defense* in this Franciscan library. It was quite obviously the same codex. But Argellati reads the date of the *Defense* as 1464. The librarian at Pavia, Dr. A. Lo Vasco, has written me confirming the data of 1454 as given in the *Inventario dei Manoscritti della R. Biblioteca Universitaria di Pavia*, Milan, 1894, p. 173.

53 *Questio de caliditate*, Milan, 1474, f. 9v, c. 2.

54 Venice, San Marco, VI, 105, f. 14r, c. 2 "... quemadmodum ego etiam probavi in tractatu quem primo edidi de reactione et in tractatu quem secundo edidi in confirmationem dictorum in primo et confutationem argumentorum que ... Gaietanus contra opinionem meam nuper in tractatu suo formavit."

55 Unfortunately, not long after photographs of this codex had been ordered, Italy went to war, making their delivery somewhat dubious. However, this work will be made the subject of a special article should the photographs arrive later.

56 The titles of the chapters of the *Liber calculationum* have already been give by Duhem (*Études*, III, p. 477) from the edition of Pavia, 1498 and by Thorndike (*A History of Magic and Experimental Science*, III, p. 376,

in the *Inventario dei Manoscritti della R. Biblioteca Universitaria di Pavia* (p. 172) is as follows:

c. 1	Marliani Joh.	De intensione et remissione
c. 4v	Eiusdem	De intensione et remissione in difformibus.
c. 9	Eiusdem	De intensione et remissione elementorum habentium duas proprias (? thus the catalogue reads, but it should be *contrarias* as it is in Swineshead's treatise) qualitates.
c. 28	Eiusdem	Capitulum septimum de reactione
c. 34v	Eiusdem	De potentia rei (Il principio del trattato è indicato da una nota a piedi della c. 35)
c. 36	Eiusdem	De difficultate actionis
c. 39	Eiusdem	De maximo et minimo
c. 41v	Eiusdem	De terra taliter descendente
c. 52v	Eiusdem	De luminosis (senza titolo)
c. 76	Eiusdem	Regulae de motu locali (c. s.)

In his *Probatio cuiusdam sententie Calculatoris de motu locali,* which is itself a discussion of a statement made in the last chapter *(Regulae de motu locali)* of the *Liber calculationum* and possibly has some relation to the last chapter of the *Tractatus physici,* Marliani speaks of having already annotated a treatise *De difformibus* of the Calculator.[57] This is probably a reference to the second chapter of the *Tractatus physici.* A final point which convinces us that Marliani's work is a commentary on the *Liber calculationum* is the title of the fourth chapter of the *Tractatus physici,* " Capitulum septimum de reactione." Swineshead's discussion of reaction is the seventh chapter in the editio princeps of the *Liber calculationum.*

The *Tractatus physici* may possibly be the work cited by Marliani as his *Liber conclusionum diversarum.* We know that conclusions 58-61 (and possibly 58-76) of this latter work concern local motion,[58] as does the last chapter of the *Tractatus physici.* Likewise we are told that the question *de potentia rei*

note 19 and *passim*) from a manuscript and the editio princeps of Padua, ca. 1477.

57 San Marco VI, 105, f. 8r, c. 1.

58 *De proportione motuum in velocitate,* Pavia, D. Confaloneriis, 1482, ff. 3v, c. 1, 5r, c. 2.

was discussed in the *Liber conclusionum diversarum*.[59] The fifth chapter of the *Tractatus physici* concerns the same question. The identity of these two works can be finally determined when the Pavian codex has been read, since we know that conclusions 58-61 of the *Liber conclusionum diversarum* were the questions argued with Philip Adiuta of Venice [60] and we are in possession of those questions. A simple comparison of them with the last chapter of the *Tractatus physici* should be sufficient.

There is no date in the Pavian codex to throw light on when the *Tractatus physici* were composed. However, we do know that if Marliani's annotation of the Calculator's *De difformibus* is the second chapter of the *Tractatus physici,* as it seems to be, then the latter must have been written before 1460, for the former was mentioned in the *Probatio cuiusdam sententie Calculatoris,* written in that year. The fact that the *Probatio* is itself a commentary on a passage from the *Regulae de motu locali* of Swineshead suggests that the last part of the *Tractatus physici,* entitled *Regulae de motu locali,* was composed about the same time, or in 1460.

D. ANNOTATIONES IN LIBRUM DE INSTANTI PETRI MANTUANI [61]

No edition, manuscript or printed, of such a work has been found. Marliani, however, mentions it twice, first in his *Probatio cuiusdam sententie Calculatoris* [62] and then later in the *Questio de proportione motuum in velocitate.*[63] It was written prior to 1460, the date of its first citation.

---

59 *De caliditate corporum humanorum*, Milan, 1474, f. 9r, c. 2.

60 See Chapter VI *infra*, note 31 and text.

61 Peter of Mantua (Piedro da Mantoa) was a professor of natural and moral philosophy at Bologna, 1392-99. See Umberto Dallari, *I rotuli dei lettori legisti e artisti dello Studio Bolognese dal 1384 al 1799*, IV, Bologna, 1924, pp. 17, 19, 22, 23. His most popular work was a *Logica*, of which there are a number of manuscripts and editions. For his *De instanti* see Chapter V *infra*, note 9.

62 San Marco VI, 105, f. 8r, c. 1.

63 Edition, Pavia, 1482, f. 26v, c. 1.

E. PROBATIO CUIUSDAM SENTENTIE CALCULATORIS
DE MOTU LOCALI

This short, but important work is contained in manuscript in the San Marco codex VI, 105, ff. 8r-12r and in print in the Marliani incunabulum *Disputatio* etc. (See section M *infra*), ff. 19r-25r. It is dated in both editions 1460.[64] San Marco VI, 105 was used in preparing Chapter V.

F. DISPUTATIO CUM JOANNE DE ARCULIS

The editions of the *Disputatio* are the same as those of the *Probatio,* codex San Marco VI, 105, ff. 12-23 and the incunabulum *Disputatio* etc. (See section M), ff. 1-19. The incunabulum gives the date of composition as 1461. Chapter III was written on the basis of the manuscript.

The *Disputatio* is largely devoted to the cooling of heated water. However, it is also concerned briefly with two other questions, which we have omitted from our study: (1) the heating or fever of the body resulting from convulsions or spasms, and (2) the digestion and evacuation of crude matter.[65]

G. DIFFICULTATES MISSE PHILIPPO ADIUTE VENETO [66]

The only copy of the *Difficultates* is in the often cited incunabulum *Disputatio* etc. (See section M), ff. 69v-80v. The date of their composition is not given, but we know that they were written after the *Liber conclusionum diversarum* (date unknown), since they are conclusions 58-61 of the latter work, and before the *Questio de proportione motuum in velocitate* (1464), in which they are mentioned as having been written previously.[67]

64 Valentinelli has read the date in the San Marco codex as August 10, 1460; while the incunabulum has August 20, 1460. Although the figure in the manuscript is not easy to read, Valentinelli's reading appears to be correct.

65 San Marco VI, 105, f. 12r, c. 2.

66 I have been unable to identify Philip other than by his connection with Marliani.

67 See Chapter VI *infra*, notes 31, 32, and text.

H. ALGORISMUS DE MINUTIIS (ET ALGEBRA)

There are two manuscripts of Marliani's *Minutie:* Paris, BN Nouvelles acquisitions latines no. 761, pp. 1-14 and Milan, Biblioteca Ambrosiana, 203 Inf., ff. 5r-9v.[68] There is a possibility that the *Minutie* is a part of his *Algebra,* which he cites in the *Questio de proportione.*[69] There is apparently no extant copy of the *Algebra.*[70]

We have no way of determining the date of the *Algorismus de minutiis,* but if it was written about the same time as the *Algebra,* either as a part of that work or separately, it would be anterior to the *Questio de proportione* (1464), since the *Algebra* is mentioned in the latter work.

I. QUESTIO DE PROPORTIONE MOTUUM IN VELOCITATE

There are three manuscript copies of the *Questio de proportione:* codex 160 of the San Marco manuscripts at Florence (Biblioteca Nazionali, convent. Soppr. J. VIII, 29),[71] Ms.

68 The Paris manuscript was used in preparing Chapter VII, although it seems to be incomplete. Repeated inquiries of the Biblioteca Ambrosiana as to the *capitula* and the *explicit* of its copy have brought no response.

69 *De proportione*, Pavia, 1482, f. 24r, c. 2 " Habes pro parte in expositione quam edidi in libellum (de) Algebra...ubi demonstrative (probavi) causam multorum tantum practice dictorum in illo libello circa modum addendi, multiplicandi, et dividendi et diminutum per diminutum et additum per diminutum et ignotum per ignotum... "

70 The treatise that follows the *Algorismus de minutiis* in the Paris manuscript turns out to be a Latin translation by Gerard of Cremona of the algebra of al-Khwarizmi. The proximity of this work to the *Minutie* apparently fooled Johannes de Sitonis in the eighteenth century. He seems to have examined the same manuscript when it was at Milan. He believed that the Gerard translation was an original work by Marliani. See B. Corte, *Notizie istoriche* etc., Milan, 1718, p. 30. This copy of Gerard's translation might have been identified earlier had not a portion of the first part been missing, so that its *incipit* was not the one commonly cited for the Gerard translation. This translation has been published by G. Libri, *Histoire des sciences mathématiques en Italie*, Vol. I, Paris, 1838, pp. 253-297.

71 A. Björnbo, "Die mathematischen S. Marcohandschriften in Florenz," *Bibliotheca Mathematica*, 3rd series, Vol. XII (1912), p. 99.

Cod. Lat. no. 2225 in the Vatican at Rome,[72] and the much later seventeenth century Harleian manuscript no. 3833.[73] In addition, it was printed by D. de Confaloneriis at Pavia in 1482 (See section M *infra*). The incunabulum will be cited in Chapter VI.

The Florentine manuscript gives us what appears to be the correct date of composition, August 27, 1464. The printed edition has no date; while the Vatican [74] and Harleian manuscripts bear the incredible date of 1407. These last two manuscripts speak of the treatise having been finished at Pavia. The Florentine codex, however, remarks that it was completed at Milan. Since the latter manuscript is the more trustworthy with regard to the date of composition, it is probably correct on this point also. Finished as it was on August 27, Marliani had no doubt come home to Milan from Pavia during the summer recess of the university.

Duhem makes the mistake of thinking that this work was printed twice,[75] once alone in 1482 and then again in the undated incunabulum *Disputatio* etc. (See section M). He deduced this from the *explicit* of the latter volume, which includes among other titles *Questio de proportionibus*.[76] Actually, however, this reference is not to any work in that particular volume, but to the separately printed *De proportione*, which was meant to be the first volume of Marliani's collected works.

72 L. Thorndike, *A Catalogue of Incipits of Medieval Scientific Writings in Latin*, Cambridge, Mass., 1937, c. 362. This manuscript falls outside the codices included in the published catalogues of the Latin manuscripts in the Vatican library. The date of 1444 given by Thorndike is that of another work in the codex. See note 74 *infra* for the date of Marliani's work.

73 *A Catalogue of the Harleian Collection of Manuscripts in the British Museum* (London), 1808, Vol. III, p. 85.

74 Vat. cod. lat. 2225, f. 37v, c. 1. " Hec ita scibere finivi papie die quartodecimo februarii hora 17ª millesimo quatricentessimo septimo etc. Amen."

75 *Études*, III, p. 499.

76 See note 81 *infra*.

J. QUESTIO DE CALIDITATE CORPORUM HUMANORUM TEMPORE HYEMIS ET ESTATIS ET DE ANTIPERISTASI

There are no extant manuscript copies of the *Questio de caliditate,* but two printed editions: Milan, Antonius Zarotus, 1474 and Venice, Bonetus Locatellus for Octavianus Scotus, 1501. There is no striking difference in the two editions. The *editio princeps* is cited in Chapter V.

The *explicit* of the *Questio de caliditate* informs us that it was completed on the eighteenth of November, 1472, during the reign of Galeazzo (Maria Sforza). As far as we know, this is the last of Marliani's works.

K. DE FEBRIBUS OMNIBUS COGNOSCENDIS ET CURANDIS
(Doubtful)

I have been unable to find any manuscript or printed edition of such a treatise. Nor is there any contemporary reference to it by Marliani or anyone else. Its first attribution to Marliani was made by Girolamo Ghilini in 1647.[77] It is quite possible that this attribution is false, since Ghilini is generally misinformed on Marliani. He makes the statement that Marliani flourished in 1430, and that he served as physician to Galeazzo Visconti and to his son Gian-Galeazzo. Ghilini has confused our Giovanni with a much earlier one, who was a ducal physician to the men mentioned and who began his teaching career at Pavia in 1374.[78]

L. EXPOSITIONES SUPER XXII FEN TERTII CANONIS AVICENNE ET DE URINIS ET MEDICAMENTIS (Spurious)

These *Expositiones* and also the accompanying treatises have been falsely attributed to Marliani by Argellati.[79] The Milanese historian cites the title as *Expositiones super XXII*

---

77 Ghilini, *Teatro d'huomini letterati,* Venice, 1647, 2 volumes in one, Vol. II, p. 132.

78 *Memorie,* I, p. 100; Parodius, *op. cit.,* p. 139. Ghilini's reference to such a work on fevers was taken up by Picinelli (Filippo), *Ateneo dei letterati Milanesi,* Milan, 1670, and from Picinelli by Argellati, *op. cit.,* II, 1, 868.

79 Argellati, *op. cit.,* II, 1, 868.

Fen Tertii Canonis Avicenne, Mediolani, apud Jacobum de Sancto Nazario, 1594 in fol; De urinis, extat in eodem; De medicamentis, extat ibidem. We are struck first by the fact that the date of publication is erroneous. Jacobus de Sancto Nazario was a Milanese printer of the latter part of the fifteenth century, and not the sixteenth. Next, upon examining the various editions printed by Jacobus, we find the following: Gian Matteo Ferrari da Grado, *Expositiones super 22 fen tertii Canonis Avicenne,* Mediolani, Jacobus de Sancto Nazario, 1494. There were two editions of this in 1494, the first of which contains a *Tractatus de urinis.* It is quite evident from the identity of the title, printer, etc. of the work attributed to Marliani with those of the treatise by Ferrari da Grado that Argellati has somehow become confused as to the author of the *Expositiones.* A careful examination of the work in question reveals nothing that refers to Marliani.

M. OPERA OMNIA

During the last years of Marliani's life when he was serving as ducal physician, a project was undertaken to collect and publish his various works. The printer was Damianus de Confaloneriis of Pavia. His name appears in the colophon of the *Questio de proportione,* printed in 1482. This was the first volume of the projected *Opera omnia.* This is deduced from the fact that in the *explicit* of the second volume printed by D. de Confaloneriis [80] the title *Questio de proportionibus* is listed first among Marliani's works; [81] yet no such treatise is

80 Although this volume has no colophon which dates it or gives the name of the printer, there can be no doubt that it was printed by Confaloneriis, for the type, its spacing by line and column, the paper, etc. are exactly alike in both volumes.

81 The *explicit* of this second volume is as follows (f. 8ov) : " Expliciunt opera subtilissima clarissimi artium ac medicine doctoris Johannis Marliani ducalis physici primi etatis omnium physicorum principis, scilicet Questio de proportionibus. De reductione aque calide. Probatio cuiusdam sententie Calculatoris de motu locali. Uterque tractatus de reactione cum tractatu Gaietani. Conclusiones quedam cum responsionibus ac replicationibus domni Philippi Adiute. Laus Deo."

included in that volume. The reference is obviously to the separately printed volume. The title of the second volume begins: *Clarissimi philosophi et medici Johannis Marliani Mediolanensis disputatio cum praestantissimo medico Magistro Johanne de Arculis in diversis materiis ad philosophiam et ad utramque partem medicine pertinentibus etc.*

Strictly speaking, these two volumes do not represent the *Opera omnia* of Marliani, since they do not include the *Questio de caliditate,* which was printed eight years earlier, the *Tractatus physici,* the annotations on Peter of Mantua's *De instanti,* the *Algebra,* or the *Algorismus de minutiis.*

CHAPTER II
REACTION: EARLY SPECULATIONS ON HEAT

WE first turn our attention to one of the problems of heat action that are so common in the fourteenth and fifteenth centuries, namely reaction. By reaction in this case we mean the cooling of a hot agent produced by a cold patient when the former is heating the latter.

The study of heat can be considered from two points of view, first the descriptive, and second the theoretical or causal. As Professor Boutaric, the noted French physicist, has put it, the first approach permits us to calculate *combien des choses;* the second gives us a basis for asking the *comment des choses.*[1] There has been a good deal more research on the development of the causal approach in antiquity and the middle ages than of the descriptive.[2] It would be wrong, however, to infer from this lack of research that there was no developed system involving the temperature concept. On the contrary, a system of four orders or degrees of hot and cold dates back at least to Galen, who used such a system to classify medicinal simples.[3] While this classification seems to have been based on a misconception of the qualities of hot and cold as fixed properties

---

1 A. Boutaric, *La chaleur et le froid*, Paris, 1927, p. 272.

2 Papers such as the following have been devoted to various theories of heat: F. Cajori, "On the History of Caloric," *Isis*, vol. 4 (1921-1922), pp. 483 ff.; L. Bauer, "Die Wärmetheorie des Robert Grosseteste" in *Die Philosophie des Robert Grosseteste*, Münster, 1917 (*Beiträge zur Geschichte der Philosophie des Mittelalters*, Band 18, Heft 4-6, 1917), pp. 157-163; A. Mitterer, "Der Wärmebegriff des hl. Thomas etc.," *Beit. z. Gesch. d. Phil. d. Mittel.*, Supplementband III, 1 Halbband, Münster, 1935, pp. 720-734. K. Meyer makes slight allusion to the second approach among Galen and the Islamic doctors in her *Die Entwickelung des Temperaturebegriffs im Laufe der Zeiten*, Braunschweig, 1913.

3 Galen treats the whole problem of the temperature concept in the third book of his *De simplicium medicamentorum temperamentis ac facultatibus*. It is in the thirteenth chapter of that book (Edition, Kühn, Vol. XI, Leipzig, 1826, p. 570 ff.) that he proposes his system of four degrees.

of the simples, it still may be called a temperature scale since it attempted to express numerically what appeared to the senses as different temperatures.

Galen's scale of degrees was taken up and elaborated upon by the Islamic physicians, the most noteworthy example of whom was Alkindi. In the latter's *De investigandis compositarum medicinarum gradibus* he shows how to calculate the degrees of medicinal compounds on the basis of the degrees of their component parts. In doing so, he demonstrates some familiarity with the fundamental distinction between intensity and quantity of hot and cold,[4] which the fourteenth and fifteenth century philosophers were to apply to more general physical problems. After its development in Islam, Galen's numerical method of describing apparent temperature differences passed into the West, possibly first by way of Constantine

---

4 *De investigandis* etc. published with Petrus de Abano, *Supplementum in secundum librum compendii secretorum medicinae Ioannis Mesues* . . . , Venetiis, Apud Iuntas, 1581, ff. 269r-273r. Alkindi points out that a compound is in equality if its frigidity is equal to its calidity. It is hot in the first degree if its calidity is twice its frigidity. When it is hot in the second degree, however, its calidity is four times its frigidity; and in the third degree eight times, etc. (i. e. the relation of calidity to frigidity proceedes by the series 1, 2, 4, 8, 16; rather than that of Galen, which is simply 1, 2, 3, 4. See f. 271r, c. 1). A typical problem in which Alkindi makes use of both the degree of intensity of a medicine and the quantity of the medicine at that intensity in determining the ultimate degree of intensity in a compound is the following (f. 272r, c. 1) : One unit weight of *mastix*, which is hot in the second degree, is mixed with two units of *cardamomum*, hot in the first degree. Then the calidity (in relation to its frigidity) of the *mastix*, which is 4, is multiplied by its quantity or weight, which is one. To this is added the product of the calidity of the *cardamomum*, which is 2, and its weight, which is 2. On the basis of this calculation the final calidity of the compound (in relation to its frigidity) is 8. The final frigidity is found similarly, and is 3. We know, however, that when the proportion of calidity to frigidity is 2 : 1, the compound is in the first degree. When it is 4 : 1, the compound is in the second degree. Therefore, we conclude that the compound of *mastix* and *cardamonum*, which has a proportion of 8 : 3, is hot somewhere between the first and second degree.

Curt Lantzsch has published a dissertation on this treatise : *Alkindi und seine Schrift de medicinarum compositarum gradibus*, Leipzig, 1920. I have been unable to locate this study. It seems to have been written under the direction of Sudhoff.

the African, who composed or translated a *Liber graduum simplicium*.[5] This work was followed by many other translations and compositions in the next three centuries.[6]

We are interested in this chapter, however, not so much in the medicinal degrees, but rather how the concepts of intensity and quantity of hot and cold were applied to physical heat problems (particularly reaction) by the schoolmen of the late middle ages, and more specifically by Giovanni Marliani.

Let us glance for a moment at the descriptive system that Marliani employs in his discussions of reaction and the reduction of hot water. He is interested in a quantity *(quantitas, multitudo)* of heat or cold because it is with this instrument that he can decide the possibility and rate of a particular heat action, by determining the powers of the agent and the patient for acting and resisting. He uses the general terminology of the Aristotelian conception of an action. An action takes place as the result of a contrariety of qualities; in the case of heat actions, hot and cold. The more powerful of these is spoken of as the agent, the weaker as the patient. The agent acts and the patient resists. Since action is a special case of motion (sc. motion of alteration), the general Peripatetic law of motion can be applied, so that an action is described by the proportion

5 In Constantine's *Opera*, Basileae apud Henricum Petrum [1536], p. 342 ff. He presents in the introduction (p. 343) a scale of four degrees for medicinal simples. These degrees are compared with the temperature of the human body. The first degree of heat is less than that of the human *complexio*; the second is equivalent to it; while the third and fourth are successively above the natural temperature. Actually in describing the degrees of the various medicines, he uses twelve different classifications, since medicines are described as being hot toward the beginning, middle, or end of any degree. *Cf.* L. Thorndike, *A History of Magic and Experimental Science*, I, pp. 750-751.

6 In their *Catalogue of Incipits of Mediaeval Scientific Writings in Latin*, Cambridge, 1937, Thorndike and Kibre have listed in the index (under "degree") a good number of treatises on medicinal degrees. Included there are several anonymous *De gradibus medicinarum*, as well as works by Maimonides, Arnold of Villanova (a commentary on Alkindi), Bernard Gordon, Gentile da Foligno, Thomas de Garbo, Guido, Jordanus de Turre, Laurentius Maiolus, etc.

of the active to the resistive power. Consequently, when Marliani wished to know whether an action would take place or the rate of that action, he first computed the active and resistive powers on the basis of the quantities of heat and cold. Quantity of heat and cold was considered as the extension of a certain degree of intensity of heat or cold. This quantity varied not only with volume, but also with density (although in most of the problems posited, the density of the agent and that of the patient are the same, and hence drop out of the proportion).[7] The power, as Marliani calculated it, can be represented by the formula $P = VT$ (or DVT, including density) where V is the volume, and T is the intensity expressed in degrees.

Before proceeding further with the Peripatetic law, we must examine Marliani's system or scale of degrees employed to represent intensity. Every body, according to the theory of the schoolmen, had limits of heat or cold reception beyond which it could be neither intensively hotter nor colder. These limits were spoken of as the "highest possible degree of calidity" (summus gradus caliditatis) and the "highest possible degree of frigidity" (summus gradus frigiditatis). When a body had not attained either summus degree, it was considered in a state or condition of hot and cold, which was a balance or equilibrium of a definite degree of heat (calidity) and a definite degree of cold (frigidity). Every alteration (latitude) toward the highest degree of calidity was accompanied by a corresponding decrease of its coextensive frigidity toward its lowest degree (zero). A scale of eight degrees of calidity and frigidity was employed by Marliani, so that we can represent any degree of frigidity and the calidity coextensive with it by the formula $F° = 8 - C°$. When $F°$ is 8, its highest degree, $C°$ is zero.[8]

7 The many problems posed in this and the next chapter amply demonstrate Marliani's theory of quantity of heat. See especially the cases cited in Chapter III, notes 27, 36, and the text.

8 Such a scale does not mean that Marliani questioned the Peripatetic

Assuming the preceding conceptions of active and resistive powers and intensity, the Peripatetic law which describes the rate of a heat action of a body *a* on a body *b* declares that the rate at any moment is dependent on the ratio of the heating

doctrine that contrary qualities can not exist together in the same subject. No, the schoolmen made a distinction between contraries and qualities of the same species as contraries. The qualities were considered contrary when they would act upon each other, and of the same species as contraries when they would not actually act themselves upon each other, but would only help other contraries to act (see Chapter III *infra*, note 45 and text). Qualities of the same species as contraries could exist simultaneously in the same subject. Let me explain further. If eight were the *summus* degree of heat in the subject, then it would be possible, if in that subject the *summus gradus* were remitted (i. e. decreased) to six degrees, for two degrees of frigidity to exist coextensively with the six. In fact it would be necessary, for the remission of calidity could not take place without an increase of the frigidity. The calidity and frigidity would not be contraries (i. e. act on each other) unless their sum were greater than the *summus* degree of calidity; that is, for example, unless the decreased calidity was six, while the frigidity was greater than two.

This doctrine originated apparently with Jean Buridan (dead after 1358). In his *Questiones super octo physicorum libros Arist.*, Parhisiis, 1509, lib. III, quest. III, f. 44r he notes that as long as the qualities are a part of the *summi gradus*, they can be coextensive. He uses as an example a scale of ten degrees (instead of Marliani's eight), and calls the equilibrium or balance of the two qualities, the *forma media perfecta*.

Duhem has noted this doctrine of Buridan, but only for the cases where bodies are unequally hot in diverse points (*Études sur Léonard de Vinci*, III, Paris, 1913, p. 402). It extends also to bodies uniformly hot (but on the way to cooling). Marliani always decides whether these coextensive qualities are uniformly or difformly spread throughout the body. Duhem's conclusion concerning this theory of Buridan is particularly interesting when we consider the problems in which Marliani constantly uses the same scale of eight degrees with the frigidity a complement of the calidity; he says: " Cette opinion qu'il n'eût pas fallu modifier beaucoup pour transformer en celle-ci: L'intensité du froid n'est que l'intensité de la chaleur *changée de signe*, cette opinion, disons-nous, attira vivement l'attention des scholastiques de Paris."

Marliani depended on Jacobus de Forlivio (Giacomo della Torre, dead in 1414) and "many others" for this doctrine; so he tells us in his *Reductio aque calide* (See Chapter III *infra*, note 45 and text). Giacomo repeats the doctrine of Buridan very carefully in his *De intensione et remissione* (First edition, Pavia? ca. 1488, ff. 14r-20v; edition with Walter of Burley *De intensione*, Venetiis, 1496, ff. 20v-23r) in the nine conclusions discussed there.

power of a to the cooling power of b (i. e. if a is heating b).[9] Then, if the temperature of a is such that it will act on b to heat it, the rate of heating depends on the quantities of heat and cold of the agent and the patient, since, as we have seen above, the powers are dependent on the quantities of heat and cold. According to the conventional interpretation of the law, the rate must arise from a proportion of greater inequality (i. e. $a/b > 1$).[10] Otherwise there would be no rate at all.[11]

So much for Marliani's descriptive system and its implications. Let us now consider the problem, whether an agent is affected by the patient on which it acts.

Doubts of the possibility of reaction are strictly medieval. Aristotle and the early commentators, as Pietro Pomponazzi tells us in his sixteenth century *De reactione*,[12] did not question reaction. Aristotle made one important distinction, however. Reaction does not necessarily take place from the action of a first agent, but it always results in the action of a last agent. The example he cited was that of a doctor who gives food or wine as a medicament. He is the first agent of the healing; while the wine or food, which actually performs the cure, is the last agent, and so suffers reaction. The last agent must always be in contact with the patient, but the first need not be. " Those active powers, then, whose forms are not embodied in matter are unaffected, but those whose forms are in matter are such as to be affected in acting." [13]

9 For confirmation, I point to the many numerical cases cited in this and the following chapter; see, for instance, the case used in the sixth premise on reaction, note 67 *infra*.

10 For an explanation of the medieval terminology of proportions, see Chapter VI *infra*, note 14 and *passim*.

11 For a general criticism of the Peripatetic law in its various forms, see Chapter VI *infra*.

12 *Tractatus acutissimi, utilissimi, et mere peripatetici, De intensione et remissione formarum . . . De reactione . . .* , Venetiis, Heirs of Octavianus Scotus, 1525, f. 21r, c. 1.

13 *Generation and Corruption*, I, 7, 324b.

It is among the fourteenth century commentators that difficulties in the accepted version of action and reaction appear for the first time. These commentators discussing reaction were of three schools. The first was the Oxford or English school, represented by Walter of Burley (dead after 1343),[14] Richard Swineshead (Suiseth), known as the Calculator, (fl. 1350?),[15] William Heytesbury (Hentisbury), who was mentioned variously at Oxford from 1330 to 1371,[16] and the anonymous *auctor de sex inconvenientibus,* who, we know, flourished sometime after the composition of Heytesbury's *De motu* (i.e. his *Tria predicamenta de motu*).[17] The second was the Parisian school, including two Germans, Albert of Saxony, who was at Paris from 1351 to ca. 1362 and was first rector of Vienna in 1365,[18] and Marsilius of Inghen, who taught at Paris from 1362 until 1382 and became first rector at Heidelberg in 1386.[19] And, finally, there was the Paduan or Italian school with Giovanni Casali, who composed his *De velocitate motus alterationis* in 1346,[20] and was Papal Nuncio to Sicily in 1375;[21] Jacobus de Forlivio (Giacomo della Torre), who practiced and taught medicine at Padua on occasions from 1402 to 1414 (the records give February 1413 as the time of his death, but since their new year began at Easter, in our

14 F. Ueberweg, *Grundriss der Geschichte der Philosophie,* Vol. II, B. Geyer, *Die patristische und scholastische Philosophie,* Berlin, 1928, p. 621.

15 See appendix.

16 P. Duhem, *Études sur Léonard de Vinci,* 3rd series, Paris, 1913, pp. 406-7.

17 *Ibid.,* p. 423.

18 Geyer, *op. cit.,* p. 600.

19 P. Duhem, *Le Système du monde,* IV, Paris, 1916, p. 165.

20 This date is given by P. A. Lopez from the *explicit* of a manuscript in the Bibliotheca Riccardiana at Florence, see " Descriptio codicum Franciscanorum Bibliothecae Riccardianae Florentinae ", in *Archivum Franciscanum Historicum,* Vol. I (1908), p. 116. I have used, however, the printed edition with the *Questio de modalibus* of Bassianus Politus, Venetiis, Bonetus Locatellus, 1505.

21 G. Sbaraglia, *Supplementum et Castigatio ad Scriptores trium ordinum S. Francisci...* II, Rome, 1921, p. 52.

reckoning February would occur in 1414) ;[22] Paul of Venice (Paolo Nicoletti), a teacher at Padua in 1408, in Siena in 1420, and dead in 1429;[23]Gaetan of Tiene, who also taught at Padua from at least 1424 and died after an extremely long career in 1465 at the age of 78;[24] and last of all, Marliani, who of course did not teach at Padua, but at Pavia.

All of these men accepted action with reaction between bodies where the qualities were difformly (i.e. unevenly) distributed, for then it would be a simple matter for the strongest part of the patient to react on the weakest part of the agent. All but two of them, Giovanni Casali and Richard Swineshead, while recognizing the difficulties to be solved, also accepted reaction among uniform qualities of the same contrariety (i.e. action and reaction between a quality and its contrary where both are evenly distributed).

First we shall examine some of the difficulties to reaction of this latter type, then some affirmative arguments which seem to show it as an absolute necessity, and finally the various solutions with their criticism by Marliani.

The most important objections to reaction are:

(1) An agent would act on a passive quality more intense than itself.[25] We might paraphrase this by saying that, other things being equal, a body which is less cold than another is hot acts on that body to cool it. This objection is precisely the same as that which declares that, assuming reaction, an action would proceed continually from a proportion of lesser inequality (i.e. an action can result even when the active power is less than the resistive power).[26] (2) If reaction is assumed to take place,

22 Duhem, Études, III, pp. 485-6.

23 Duhem, Le Système du monde, IV, pp. 280-283.

24 Ibid., IV, p. 301.

25 Giovanni Casali, Questio ... de velocitate motus alterationis with the Questio de modalibus of Bass(i)anus Politus, Venetiis, Bonetus Locatellus, 1505, f. 65v, c. 2.

26 Walter of Burley, In octo volumina ... Aristotelis de physico auditu ..., Padue, 1476, III Bk., f. 69v.; Jacobus de Forlivio, Tractatus de intensione et

no point could be found at which it takes place.[27] This is based
on the assumption that the part of the agent which would be
affected *(pars repassa)* would always be in a proportion of
lesser inequality to the remainder of the agent, and one would
be unable to choose at any moment an affected part which
would not show a lesser proportion. (3) If reaction were accepted, there would be apparently nothing to prevent an infinity
of actions and reactions; that is, the part of the agent on which
the patient had reacted would in turn react on a part of the
patient, and so on, *ad infinitum.*[28] (4) From reaction it seems
to follow that contrary qualities are increased in intensity in
the same subject in the same time, or that the same thing is
moved with contrary motions. For example, some air which
occupies the position of a medium between a hot and a cold
body would be cooled and heated at precisely the same time.[29]

remissione with Walter of Burley's *De intensione et remissione,* Venetiis,
1496, f. 27r, c. 2 (also in the first edition of Jacobus' work, Pavia? ca. 1488,
f. 33r, c. 1); *Tractatus de sex inconvenientibus,* with the *Questio de modalibus*
of Bass(i)anus Politus, Venetiis, Bonetus Locatellus, 1505, f. 43v, c. 1;
Albert of Saxony, *In eosdem (libros de generatione) ... quaestiones,* with the
In Arist. libros de Generat. commentaria of Egidius Romanus, Venetiis,
Hieronymus Scotus, 1567, p. 307, c. 2 (quest. 16); Marsilius of Inghen, *In
eosdem (libros de Generatione) ... quaestiones,* (In the edition above with
Egidius and Albertus), quest. 19, p. 188, c. 1. Pietro Pomponazzi, *op. cit.,*
f. 21r, c. 1.

27 Giovanni Casali, *op. cit.,* f. 65v, c. 2 (2nd objection). Albert of Saxony,
op. cit., p. 308, c. 1 (6th doubt). Jacobus de Forlivio, *op. cit.,* Ed., 1496, f.
27v, c. 1; Ed. ca. 1488, f. 33v, c. 2 (5th objection). *De sex inconvenientibus*
(1505), f. 43v, c. 1 (1st inconvenience).

28 Giovanni Casali, *op. cit.,* 65v, c. 2 (3rd objection); Jacobus de Forlivio,
op. cit., Ed. 1496, f. 27v, c. 1, Ed. ca. 1488, f. 34r, c. 2 (7th objection);
Marsilius of Inghen, *op. cit.,* p. 187, c. 1 (8th objection).

29 Albert of Saxony, *op. cit.,* p. 307, c. 2 (3rd doubt); Marsilius of Inghen,
op. cit., p. 188 (4th doubt), Jacobus de Forlivio, *op. cit.,* Ed. 1496, f. 27r,
c. 1, Ed. ca. 1488, f. 33r (1st and 2nd objections); Richard Swineshead,
op. cit., f. 30r, c. 1; Gaetan de Thienis, *Recollecte super octo libros physicorum
Aristotelis,* Venetiis, Bonetus Locatellus, 1496, f. 26r, Pietro Pomponazzi,
op. cit., f. 21r, c. 1 (2nd and 3rd difficulties).

(5) No agent would be able to assimilate a patient to itself perfectly.[30]

These five chief objections and several others were not strong enough to counteract three very important arguments used to support reaction. The first of these is the argument by experiment, by which it is shown that a heating agent is invariably cooled when it heats a patient. Several examples of such an " experiment " or " experience " are given by all writers. If a warm hand holds a cold piece of fruit (pomum) or a stone, the hand is cooled while the fruit or stone is warmed. Similarly when a hot iron is plunged into cold water, the water is heated and the iron is cooled. The mixing of hot and cold water always finds the hot cooled and the cold warmed, etc.[31]

This last example of mixing hot and cold water brings us to our second affirmative argument, namely, without reaction no compound or mixture could possibly be generated out of the elements, or no third element could result from two elements.[32] In other words, if reaction were discarded, the whole Aristotelian theory of action by contrariety would be in danger.

To these two arguments Jacobus de Forlivio adds a third. Every agent would be of infinite resistance if there were no reaction. Such is obviously not the case.[33]

30 Albert of Saxony, *op. cit.*, p. 308, c. 1 (4th doubt) ; Marsilius of Inghen, *op. cit.*, p. 188 (3rd doubt) ; Jacobus de Forlivio, *op. cit.*, Ed. 1496, f. 27v, c. 1; Ed. ca. 1488, f. 34r, c. 1 (6th objection) ; Pietro Pomponazzi, *op. cit.*, f. 21r, c. 1 (4th difficulty).

31 These are only a few which are cited by the authors. See the four examples of Albert of Saxony, *op. cit.*, p. 307, c. 2, and the similar ones of Marsilius of Inghen, *op. cit.*, p. 188, c. 2 (both of these men include the three experiences noted above). *Cf.* Jacobus de Forlivio, *op. cit.*, Ed. 1496, f. 26r-v, Ed. ca. 1488, f. 30r-v.

32 Marsilius of Inghen, *op. cit.*, p. 188, c. 1 " ... quodlibet miscibilium agat et repatiatur, aliter enim ex eis non posset mixtum generari...''; Albert of Saxony, *op. cit.*, p. 307, c. 2 " Si reactio non esset possibilis, sequitur quod ex duobus elementis per eorum actionem et passionem adinvicem non posset tertium generari, set hoc est falsum... "; *cf.* Jacobus de Forlivio, *op. cit.*, Ed. 1496, f. 26r, c. 2, Ed. ca. 1488, 30r, c. 2.

33 Jacobus de Forlivio, *op. cit.*, Ed. 1496, f. 26v, c. 1, 30v, c. 2, " ... si non reageret passum in agens et cetera, sequitur quod omnes agens in certam resistentiam esset resistentie infinite, quod est impossibile... "

We have noticed that these affirmative arguments were of such an importance (especially the argument by experience) that the overwhelming majority of the physicists accepted reaction. In opposing reaction, Giovanni Casali had offered difficulties, but made no attempt to solve the affirmative argument of experience.[34] The Calculator, on the other hand, in rejecting reaction between uniform qualities [35] makes an attempt to solve the experiences by reducing them to cases of actions between difform qualities (unevenly distributed qualities). In these solutions he introduces a kind of double *caloric* theory of heat and cold. For instance, the experience of a cold hand on a hot head, with the consequent heating of the hand and cooling of the head, is explained thus: small hot bodies go out from the head, and enter the pores of the hand, heating the weak bodies there. As a result of this action the hand feels hot. Conversely, cold bodies in the hand leave the hand and cool the weak bodies in the head. And so wherever there is reaction, it is a result of a similar difformity.[36]

34 Giovanni Casali, *op. cit.*, f. 6or, c. 1, 65v, c. 2. He expresses his opinion thus: " Intelligo istam conclusionem (de reactione) in sensu composito sic: non est possibile aliquam qualitatem reagere in qualitatem que agit in eam."

35 Richard Swineshead, *op. cit.*, f. 34v, c. 1 " Immo summarie dicitur quod inter uniformia contraria, loquendo de uniformi ita quod omnes partes equales equaliter contineant de forma sive illa uniformia sint equalia in potentia precise sive inequalia, non est reactio, loquendo de reactione duos motus scilicet contrarios includente ... tamen difformi bene potest esse actio cum reactione. Vel hoc secundum diversas partes tam inter equalia potentie quam inequalia."

36 *Ibid.*, f. 34r, c. 1 " Immo omnino est dicendum ad quantum experimentum quod manus frigida apposita capiti calido dolenti manus calefit et caput frigefit quia parva corpora calida exeunt a capite et ingrediuntur poros manus et calefaciunt corpora debilia inclusa in manu, ratione cuius actione sentitur manus calida. Et econtra corpora frigida in manu sive a manu exeuntia agunt frigiditatem in corpora debilia in capite vel in carne inclusa. Etiam solvuntur omnia experimenta que de actione naturali inter elementa allegantur. Ubicumque enim sentimus reactionem aliquam, ibi sunt inter corpora mixta multum difformia, ratione cuius difformitatis per aliquam partem potest esse actio et per aliquam econtra reactio."

So much for the opponents of reaction. The problem facing the advocates was to provide a solution which would do away with the major objections already enumerated. Almost all of the authors posed the so-called reaction by the " affected part " *(pars repassa)*. A hot body *a* acts on a cold body *b*, and in turn is cooled by *b*. The part of *a* immediate to *b*, and therefore suffering action from *b* at any given moment, is called the affected part *c*. The remainder of *a*, as yet not reacted upon by *b*, is called the unaffected part *(pars non repassa) d*. The description of reaction in this manner is an obvious recognition of the part to part conduction of heat. What the authors must explain in such a scheme is how *d*, which is much hotter than the affected part *c*, does not immediately by the Peripatetic law heat that part; and since the relationship of *d* to *c* at any given instant after the beginning of the action must be a proportion of greater inequality, how can we conceive of any moment at which *c* will be cooled by *b*.[37]

The most obvious solution would be to abandon the Peripatetic law as descriptive of actions. This apparently occurred to an unknown few, for we find a method described but rejected by Albert of Saxony and Marsilius of Inghen which declared that although local motion must necessarily proceed from a proportion of greater inequality, it was possible for motion of alteration to arise from a proportion of lesser inequality.[38]

Several other solutions of reaction are rejected by Albert and Marsilius. One such solution does not admit actual reaction by the patient, but contends rather that the agent is affected by some outside influence: more immediately, the air, or more remotely, the heavens.[39] Another solution, refuted first by

37 Richard Swineshead, *Liber calculationum*, f. 30 ff. gives the best summary of these logical difficulties of the argument of the affected part.

38 Albert of Saxony, *op. cit.*, p. 190, c. 1; Marsilius of Inghen, *op. cit.*, p. 308, c. 1. " Minus posse agere in maius in motu alterationis ... "

39 Albert of Saxony, *op. cit.*, p. 308, c. 2, " ... in omni actione agens repatitur, sed hoc non est a passo, sed a circumstante propinquo, quod est aer, vel remoto, quod est coelum." *Cf.* Marsilius of Inghen, *op. cit.*, p. 190, c. 1. Richard Swineshead uses this theory as a possible solution, *op. cit.*, f. 33v.

Marsilius, and then by Marliani in the second part of his treatise *De reactione* on the grounds that it explains only limited cases of reaction, is based on the initial premise that every agent propagates its power around itself spherically, in such a way that the power is stronger nearer to the agent, and weaker farther from it. The agent acts to assimilate a contrary according to the power which is determined by the distance from the agent. In any assimilation a kind of mixed mean of the powers of the agent and its contrary occurs; and it is this mixed mean that explains reaction.[40]

After dismissing the various solutions which prove to be inadequate, Marsilius and Albert advanced one of their own which had considerable influence in Italy. This theory provided that qualities are not equally active and resistive. A quality like calidity, for instance, is very active, yet little resistive, so that something hot remits something cold because it is more active than the other is resistive; but in turn is cooled because it is less resistive than the other is active.[41] Such a theory appears

He tells us that when a hot iron is plunged into the water, the water is not really heated by the iron. Rather, there is air or some inflammable material in the water with the iron. This material is light and seeks to ascend. As long as this hot vapor is in the water, the water appears to be hot, but actually it isn't. A similar solution is the "pore" theory (ff. 33v-34r). In the iron and mixtures or compounds there are pores filled with air or subtle vapor which receive the action from a hot agent; and an exhalation takes place as a result of the subtle vapor thus heated. The dense part of the water actually acts and is not affected; it is only the rare part (the vapors) which suffers action. Compare this with the pore theory attributed by Aristotle to Empedocles, *Generation and Corruption*, I, viii, 324b.

40 Marsilius, *op. cit.*, p. 190, c. 2; Marliani, *De reactione*, Ms. San Marco VI, 105, f. 28 ff.

41 This is based on a scheme which arranges the qualities in the following order of strength of activity: hot is strongest, then cold, then wet, and finally dry. However, the strength of resistance is precisely the reverse of this; i. e. dry is the strongest, then wet, cold, and finally hot is the weakest. See Marsilius, *op. cit.*, p. 187, notanda 4, 5, and 6, and the six conclusiones, p. 188. Also *cf.* Albert of Saxony, *op. cit.*, p. 309 "... comparando caliditas est maioris activitatis quam sit frigiditas, et est minoris resistentie quam sit activitas frigiditatis. Similiter frigiditas est minoris resistentie quam sit activitas caliditatis. Secundum hoc ergo potest dici quod in approximatione calidi ad frigidum sit actio et reactio."

in one form or another in the discussions of Jacobus de Forlivio, Paul of Venice, and is mentioned as a possible solution by Gaetan of Tiene.[42] It is refuted, however, by Marliani in the third part of his *De reactione* in discussing the opinions of Jacobus, Marsilius, and Paul. He notes that Jacobus uses it, but that he does so without proof.[43] As his principal argument, Marliani shows that by using this theory reaction could not take place in some cases where experiment proves that it does. One such case is posed. One-eighth part of cold water is mixed with one part of boiling water. Experiment demonstrates that a reaction takes place, but according to the method of Marsilius this could not be confirmed. The argument, in which Marliani uses his fundamental conception of active and resistive powers based on quantity of heat or cold as a powerful weapon against the theory of Marsilius, runs like this: (1) There is no way in which, according to the method of Marsilius, you could show that heat (calidity) is eight times as great in active power as in resistive power. Marsilius would no doubt admit this. (2) Therefore, although all cold (actually, *quelibet frigiditas*) is of a greater active power than resistive, according to Marsilius, it does not follow here that the active power of the frigidity of the cold water in the example above is greater than the resistive power of the calidity of the boiling water. In fact, the opposite is true and can be proved as follows: first of all, we can ignore Marsilius and assume that frigidity and calidity are equally active in general. Then on the basis of quantities *(multitudines)* of heat and cold, the active power of frigidity of the one-eighth part of cold water would be one-eighth of the active power of the calidity of the one part of boiling water. (In computing the powers Marliani mentions here that the ratio would be one to eight, if the one-eighth part and the whole had the same intensities, i. e. the hot water would be just as hot in the number of degrees as the cold water would be

42 Jacobus de Forlivio, *op. cit.*, f. 28v; Paul of Venice, *op. cit.*, f. 46v, c. 2; Gaetan of Tiene, *op. cit.*, f. 26r.

43 Marliani, *De reactione*, Ms. cit., f. 31v, cs. 1-2.

cold, and if they were of the same density; or the ratio of the powers would be one to eight if the ratio of the *multitudo frigiditatis* to the *multitudo caliditatis* was one to eight.) But according to Marsilius frigidity is less active than calidity. Consequently the frigidity in question is even less than one-eighth of the active power of the calidity. This would mean that the weak active power of the cold water which is less than one-eighth the active power of the calidity would have to be more powerful than the resistive power of the calidity (i.e. the resistive power of the calidity in general would have to be over eight times as weak as the active power of the calidity), which hardly seems likely. As a result, since we know by experiment that reaction takes place, and the cold water does cool the warm, we find an action taking place from a proportion of lesser inequality.[44]

Marliani in answering further the theory of Marsilius, questions the experiments that he has used to prove his contention as to the relative active and resistive strengths of the qualities. Marsilius has said, for instance, that calidity is more powerfully

44 *Ibid.*, f. 31v, c. 2 " Item hec responsio negat reactionem fieri in pluribus casibus, in quibus ipse, et alii concederent reactionem. . . . Ergo responsio non bona . . . pono quod in aqua fervente ponatur 8ª pars aque frigide et patet ex experimento quod frigida calefiet et calida frigefiet. Jacobusque et Marsilius concederent hoc. Sed ergo probo secundum datam responsionem, quod non fiat ibi reactio quia non est verisimile nullamque aduceres rationem quod caliditas in 8^{lo} maioris potentie active sit quam resistive. Ergo quelibet frigiditas secundum Marsilium maioris sit potentie active quam resistive, non sequitur tamen, quod activitas frigiditatis dicte aque frigide sit maior resistentia caliditatis aque calide. Ymmo magis sequitur oppositum. Si enim frigiditas tantum esset activa, quantum caliditas, ceteris paribus, frigiditas huius aque esset in 8^{lo} minoris potentie active quam caliditas aque ferventis, si caliditas non sit remissior frigiditate et sit paritas in dempsitate, aut saltem in multitudine ita ut multitudo frigiditatis sit 8ª pars multitudinis caliditatis, loquendo secundum ordinem nature consuetum. Sed frigiditas brevioris est potentie in agendo quam caliditas, ceteris paribus maxime secundum ipsos. Ergo: frigiditas aque frigide, plusquam in 8^{lo} minoris potentie active quam caliditas aque calide et illa caliditas aque calide non est in 8^{lo} minoris potentie resistive quam active. Ergo activitas frigiditatis aque dicte frigide super caliditatem dicte aque ferventis est proportio brevioris inequalitatis. Ergo non aget talis frigiditas in caliditatem."

active than frigidity, and cited as proof the fact that one can put his hand in cold water for a long time without being affected, but that if he puts it in hot water, he feels the effect immediately.[45] Marliani answers that this has nothing to do with the activity of the calidity or frigidity, but is a physiological property of human beings or living things which, you might say, have a weakness for calidity so that they actually have more heat than cold. One would expect the heat, even though it was of the same power as the cold, to affect the naturally hot human being more.[46]

The first attempt of the English school to solve the problem was that of Walter of Burley, who in a very brief discussion of it in his commentary on the *Physics* of Aristotle merely noted that as long as the agent is not placed beyond the distance through which the patient can react, or as long as the agent and patient are joined *(communicans),* action with reaction will take place. He adduces and solves almost none of the difficulties which were later so popular.[47]

The argument of the affected part *(pars repassa),* briefly noted above, plays a central role in the discussions against reaction of Richard Swineshead,[48] of William Heytesbury, and the anonymous author of the *Tractatus de sex inconvenientibus.*

William Heytesbury proposes a rather weak solution of the argument of the affected part in his *Sophismata.* If *a* is the hot agent, and *b* is the cold patient, then the whole of *a* acts on the whole of *b,* and there results an action between *a* and *b.* But action does not always take place from the proportion of a whole to a whole, but also from the proportion of a whole to

45 Marsilius, *op. cit.,* p. 187, c. 2.

46 Marliani, *De reactione,* f. 33v, c. 2 " . . . omne vivens et omnis homo maxime declinat ad calidum ita quod plus habet de caliditate quam de frigiditate. Quare cicius a calido quam a frigido pati debet, quamquam scilicet caliditatis et frigiditatis potentia in agendo equali existente . . . "

47 Walter of Burley, *op. cit.,* ff. 69v-70r.

48 See note 37 *supra.*

a part. In this way the whole of b can act on part of a adjacent to it (the so-called affected part), and the result is reaction.[49]

As Marliani points out in opposition to this opinion of Heytesbury, the fundamental question posed by the Calculator and others as to whether the unaffected part of the agent would act on the affected part, since it is hotter, is not answered by merely saying that the action of the patient b on the affected part c impedes the action of d (the unaffected part of the agent) on c, for the Calculator proved the opposite.[50]

The anonymous author of the *Six Inconveniences,* on the other hand, explains that the unaffected part of the agent does not act on the affected part because in the beginning all of the agent a is uniform, and as a result does not seek to act on any of its parts. But b the patient is dominant over the adjacent part of the agent c and so begins to act on it, and having done so it continues to act because " it is easier to continue motion than to begin it ".[51]

Marliani does not believe that this statement upon which the proof of the anonymous author is based, namely, that it is easier to continue motion than to begin it, applies to actions. In attempting to find out how it would be possible for an action to continue with less difficulty than to begin, he rules out the possibility that the action acquires an *impetus,* as in local motion and in doing so shows himself familiar with the *impetus* theory which received such an extended treatment from the Parisian scholastics of the 14th century, John Buridan, Albert of Saxony, and others.[52]

49 William of Heytesbury, *Sophismata,* with his collected works, *De sensu composito,* etc., Venetiis, 1494, f. 165v.

50 Marliani, *De reactione,* f. 35r, c. 2.

51 *Tractatus de sex inconvenientibus* (1505), 44r, c. 2 "... in principio totum a est uniforme, et per consequens non appetit agere in aliquam sui partem, sed solum in suum contrarium; et sic de b; et ... b dominatur super partem repassam puta c, ... ideo tunc b incipit agere in c, et incepta actione facilius est continuere motum quam incohare. Igitur b continue aget in c ..."

52 See Chapter VI, *infra,* note 1. Marliani treats the theory in the third part of the *De reactione* while discussing the aforementioned theory of Heytesbury. See *Ms. cit.,* ff. 36r, c. 2, 36v, c. 1.

Without trying to present the long discussion of Jacobus de Forlivio solving the various difficulties enumerated, I might note here that in addition to employing the Parisian theory of the varying strength of the active and resistive powers of the qualities, he presents a rather interesting doctrine to answer the inconvenience that a medium between a hot and cold body would be heated and cooled simultaneously if we accept reaction. Jacobus admits that cold and hot are produced in the same medium at the same time, but he points out that there is a distinction between producing frigidity and increasing its intensity, so that while frigidity is being produced simultaneously with the calidity, it is likewise at the same time being destroyed by the calidity without having its intensity increased. Therefore, actually it is only the calidity in the medium which is increasing in intensity.[53]

Among those arguments that Marliani brings against this theory of Jacobus, we notice one that questions this simultaneous production and destruction of calidity, and which deduces from Jacobus' premises that the frigidity must have a permanent nature *(natura permanentis)*, rather than the successive one which simultaneous production and destruction would demand.[54]

Perhaps the last attempt to solve reaction before Marliani was made by Gaetan of Tiene in his commentaries on the *Physics*. After presenting the solutions of Heytesbury and Marsilius, he proceeds to a final theory which he supports. We can label this the " reflexion theory ", since it has as its primary suppositions some statements borrowed from supposed principles of the reflexion of light. The suppositions are two in number : (1) There is greater activity when there is more reflexion of species. The similarity to the theory of light is stressed here when Gaetan cites in support of this supposition the fact that fire is caused by the rays of the sun when they are reflected by a concave mirror, but there is no fire when the

53 Jacobus de Forlivio, *op. cit.*, f. 28r.
54 *De reactione*, f. 29v, c. 1.

reflection is by a plane mirror.[55] Marliani believes that this is not clearly expressed. No more rays of light are absorbed by the plane than concave mirror. Therefore, all of the rays are reflected by both mirrors. What Gaetan means, according to Marliani, is that there is a greater concentration or union *(unio)* of species, rather than a greater reflexion, which produces the greater activity.[56] (2) Where there is greater contrariety, there is greater reflexion of species. In support of this contention, Gaetan mentions the Aristotelian theory which explained the production of hail in the summer by a greater contrariety in the middle region of the air. Gaetan claims that there is greater reflection of the species of frigidity in the summer than in the winter, which results in this supposed greater contrariety.[57] Marliani questions this doctrine that assumes that there is greater contrariety, and hence greater congelation, in the middle region of the air in the summer than in the winter.[58]

From these suppositions, reaction is explained by Gaetan in the following manner. The patient *b* acts on the affected part *c* by frigidity (i. e. to cool it), but the unaffected part *d* does not immediately act on *c* to heat it. The reason for this is that *b* has an instrument *(instrumentum)* more powerfully applied to *c*, than does *d* because of the greater contrariety (and greater reflexion) between *b* and *c* than between *d* and *c*. *c* is cooled and not heated, then, as a result of the greater contrariety and

55 Gaetan of Tiene, *op. cit.*, f. 26r " ... supponatur primo quod ad specierum reflexionem sequitur maior actio. Hoc patet quoniam per reflexionem radiorum solis factam a speculo concavo generatur ignis et non per (reflexionem) factam a speculo plano."

56 *De reactione*, f. 39r, " ... non enim magis penetrant radii speculum planum quam concavum, ceteris paribus, ergo omnes ita a plano sicut a concavo reflectentur ... Credo tamen intellectus eius in suppositione fuerit quod per maiorem unionem specierum, maior sit actio ... "

57 Gaetan of Tiene, *op. cit.*, f. 26r; see Aristotle, *Meteorologica*, I, 348a-b. Aristotle calls this phenomenon of greater contrariety "antiperistasis". Marliani later wrote an extended treatment of antiperistasis in his *De caliditate corporum humanorum*. See Chapter IV, Section II.

58 *De reactione*, f. 39r-v, see Chapter IV, Note 52 and discussion.

the fact that the action of one contrary impedes that of another (i. e. the fact that as long as something hot is acting, it is impossible for its contrary, cold, to act on the same subject). After a certain period of time b will be so diminished by the principal action of a (i. e. it will be warmed by a) that it will no longer have a greater contrariety with c than does the unaffected part d, and so c will begin to be warmed by d until an even *summa* intensity of calidity is reached by all parties in the action.[59]

We have seen that Marliani thinks that Gaetan's original suppositions are poorly expressed and open to question. He adds that even if they are granted, the theory is exposed to a serious objection. It does not allow reaction between bodies that are in contact *(corpora immediata)*.[60]

Up to this point we have examined the various theories of reaction which were common at the appearance of Marliani's *De reactione*.[61] This treatise has a twofold purpose, to present Marliani's solution, and to criticize all previous opinions. The first of these objectives is achieved in the first part of the tract, the last in the three remaining sections. We have already noted the most essential criticisms included by the Milanese physician in the second and third sections of the work. The fourth part attempts two things, to answer the Calculator when he said that it is not possible to have action with reaction by means of the same part (the affected part), and second to answer the

59 Gaetan of Tiene, *op. cit.*, f. 26r.

60 *De reactione*, f. 39v, c. 1. In his treatise *De reactione*, written in defense of his theory of reflexion against the attack of Marliani, Gaetan concentrates primarily on destroying the suppositions upon which Marliani bases his theories, as we shall see immediately; but he does attempt, also, to clarify his own theory. He shows in his first supposition that he accepts the theory of Marsilius concerning the relative active and resistive powers as the basis of the reflexion theory. The remaining exposition of his theory is not changed in the least from the outline in the commentaries to the *Physics*. (See the *De reactione*, which is included in the Marliani manuscript of San Marco, VI, 105, especially ff. 44v-45r).

61 See Chapter I (Section A, B) for dates of this work and of the short supplementary work defending it against the arguments of Gaetan of Tiene.

various difficulties adduced by Jacobus de Forlivio. Both of these he does by introducing some specific problems showing reaction by the affected part. He advises us here to turn to the first section if we believe that reaction cannot take place.[62]

Let us then take Marliani's advice and examine, first briefly and then in more detail, his rather curious method of explaining reaction. This method involves adjoining to a *summum* hot body *a* a body with lesser calidity (and hence some coextensive frigidity, according to the common method outlined of representing intensity), which helps it to act on *b*, a *summum* cold body, which is itself joined to a body of lesser frigidity (and hence some calidity) that helps it to react on *a*. Neither of these helping bodies, which we may call *c* and *d*, is able, however, to aid *a* and *b* to resist an action. Hence in this way we have what amounts to *c* and *a* acting on *b* alone; and *b* and *d* uniting to react on *a* alone.

It is quite obvious that without the aid of bodies *c* and *d*, there would be, according to the method of Marliani, no possible way for action and reaction to take place between two bodies, one uniformly hot of *summa* calidity and the other uniformly cold of *summa* frigidity. But this conclusion is completely adverse to experiment. Strangely enough, however, when Gaetan of Tiene launched an attack on Marliani's method in his own *De reactione,* he failed to note this fundamental criticism; instead he refuted Marliani's specific premises.

The most important thing to be proved by Marliani in order to protect the validity of his method is that the aiding bodies, *c* and *d*, help *a* and *b* to act, but do not help them to resist action. He does this by a series of premises, only a few of which need be mentioned.

The first of these suppositions takes exception to the theory of Marsilius and Albert which had recognized that a quality differed in acting and resisting power, posing in its stead that

62 *De reactione*, f. 41r, c. 2.

a quality is resistive to the same degree as it is active.[63] We can see how this supposition would fit in better with the conception of action according to a power based on quantity of heat or cold, since it must necessarily simplify and make uniform the calculation of such powers.[64]

The third supposition is the most important of all, and with it accepted, reaction as Marliani conceives it becomes understandable. No agent possessing certain properties (e.g. the property to heat) immediately and *per se* resists an action, which it itself by those properties intends to produce, and which it would produce in exterior matter, if it were more potent in resistance than the exterior matter, assuming a sufficient application had been made. Thus if any agent by such properties (e. g. to heat) has the intention of assimilating another body (and which would assimilate that body if it were strong enough), it does not immediately and *per se* resist that action (e.g. heating) when it takes place by another agent.[65] This

63 *De reactione*, f. 23r, cs. 1-2, "Accipimus itaque primo quod in corpore uniformi quantum ad qualitatem, qua immediate agit aut resistit equalis est potentia secundum illam agendi et resistendi, quod communiter sub hiis verbis dici solet, quod quelibet qualitas quantum agit, tantum resistit alteri, ne ab eo paciatur ... "

64 Gaetan protests against this supposition in his *De reactione*, Ms. cit. (in note 55), declaring that if it were so, it should be easier to dry water than to heat it; which he believes to be disproved by experiment. The confirmation of this inconvenience is based on an experiment already refuted by Marliani, namely, because hot water is more injurious to the hand than cold, heat in general is more active than cold. Marliani defends himself further in the short treatise *In defensionem libri de reactione* (San Marco, VI, 105, f. 1r).

65 *De reactione*, f. 23r, c. 2 " Tertio premitto, quod nullum agens cum talibus instrumentis et dispositionibus coniunctum, immediate et per se resistit actioni seu effectui, quem cum illis instrumentis et dispositionibus existens intendit producere, atque produceret in materiam exteriorem, maxime, si omnino taliter qualificatum esset potentius resistiva, que est in illa materia exteriore, sufficienti ad illam applicatione existente, itaque si intendat aliquod agens, cum hiis dispositionibus coniunctum, aliquod corpus assimilare, atque si potentius illo foret, manens tale, et debite sibi approximatum assimilaret, non per se et immediate resistit illi actioni seu assimilationi etiam si illa, aut talis assimiliatio ab alio agente fieret."

simply means that if a hot body a (imperfectly hot so that there is coextensive frigidity) intends to heat a cold body b, it would not resist with its coextensive frigidity the heating of b by another body c. Marliani cites a problem to illustrate this. A body a, which is one foot or unit in extension, has a heat intensity of 4 degrees with a cold intensity of 4 degrees. It touches a body b, also of one foot extension, which has a calidity of 8 degrees. Adjacent to the other side of b is a body c of four feet extension which has a frigidity of eight. Since c has a greater quantity of cold than b has quantity of heat, c acts on b. Now this third supposition declares that a does not help b with its calidity to resist this infrigidation by c because its intention is likewise to cool b (and this intention would be carried out if a were increased in power by the addition of a_1 similar in intensity of calidity and frigidity to a).[66]

One other important supposition is needed to make Marliani's case of reaction possible, namely one that allows for the active aid of c and d to a and b. The eighth is such a supposition. Although, as it is explained in the sixth (and seventh) supposition, not every part of a body difform in some quality aids every other part (irrespective of the part's intention) for acting and resisting,[67] the eighth tells us that every part in

---

66 *Ibid.*, f. 23r-v Gaetan also vigorously opposes this third supposition of Marliani, declaring that an agent immediately resists an action which it has produced or tends to produce (*De reactione*, Ms. cit., f. 43r). He cites as an example a problem involving a body a whose upper half is eight degrees of calidity and whose lower half is eight degree of frigidity; while the extension of a is one unit or foot. Body a is joined to a tepid body b which uniformly has a calidity of 4 and a frigidity of 4 and an extension of ½. The result is that the calidity of a is supposed to act on the frigidity of b, but the frigidity of a is resisting any increase in the calidity of b by acting on it to decrease it. Hence Marliani's third supposition is incorrect. But Marliani provides an answer to this problem by pointing out that the whole of a does not seek to heat or cool b, but only parts of it. Hence the problem is not applicable (*In defensionem*, Ms. cit., f. 1r).

67 *De reactione*, ff. 23v-24r. In demonstrating the sixth supposition Marliani poses a problem which shows clearly the formulae which he used to compute active and resistive powers for the Peripatetic law. This problem concerns a body a consisting of two parts b and c. c is one unit in extension and has a

both uniform and difform bodies can help every other part in producing a degree that will be less intense or equally intense as the degree of that quality which it possessed itself, or rather that it could produce itself.[68]

The distinction between the sixth and eighth suppositions must be made clear. Suppose that the calidity is difform in a body with one part 4 degrees and the other 8 degrees. If the part which is 8 degrees tends to produce in another body an intensity of 5, according to the sixth supposition, the part which is 4 cannot aid in producing the intensity of 5; but according to the eighth supposition, this part which is 4 can aid the part which is 8 in producing an intensity of less than 4.[69]

Although Marliani gives other suppositions (and several conclusions), these that we have recited are enough to show how reaction could take place. We would expect that by the third supposition, c and d would not be able to help a and b to resist, but by the eighth supposition would be able to help them to act. Hence c plus a would be great enough in power to act on b alone (unaided in resistance by d), and conversely, b plus d would be powerful enough to react on a alone (unaided in resistance by c).

With this description of Marliani's method of solving reaction, we come to the conclusion of our sketch of the various treatments of reaction. Analytically, Marliani proved quite suc-

---

calidity of 6 with a coextensive frigidity of 2; while b is three units or feet in extension, and has an intensity of calidity of 8 with no coextensive frigidity. This body a is brought in contact with a body d at its part b. This d has an extension of $2\frac{1}{2}$ units, a heat intensity of 7, and a coextensive cold intensity of 1. Marliani notes that b, c, and d are equally dense; so that the factor of density can be eliminated from our computations. Now, according to Marliani, the proportion of the power of calidity b to the power of frigidity of d is 48 to 5. This proportion was obtained by the multiplication of the respective intensities by their extensions, which confirms the law as deduced earlier in the chapter (See note 9 and text).

68 *Ibid.*, f. 24r, c. 2.

69 Gaetan again attempts to cite cases to prove the non-validity of these suppositions, *De reactione*, f. 43r-v. For Marliani's defense, see *In defensionem*, f. 2r-v.

cessful in criticizing the solutions which preceded him. Constructively, he was less successful. His treatment suffers from the same major difficulties of all the discussions, an unhesitating acceptance of the Aristotelian qualitative frame and a mistaken application of the erroneous Peripatetic law of local motion to motion of alteration (i. e. to actions). But in accepting and employing at all times a system of numerical units or degrees to represent intensities, and in distinguishing this system of units from another which described the extension or quantity of the degrees of intensity, Marliani and his predecessors laid the necessary foundation for the invention of the thermometer (i. e. they accepted and clarified the idea of degrees, which was a *sine qua non* for the invention of an instrument to measure degrees) and for the distinction between temperature and quantity of heat.

CHAPTER III

THE REDUCTION OF HOT WATER

ANOTHER problem of heat actions widely disputed by the schoolmen of the fourteenth and fifteenth centuries was that of the reduction of hot water to frigidity (i. e. the cooling of hot water). Is heated water cooled by something within the water itself, intrinsically *ex se,* or is it cooled extrinsically by the container? This is the fundamental question. That extrinsic reduction should be doubted is not remarkable when it is realized that the only thermometer of the schoolmen was their senses and that their understanding of the effects of daily and seasonal climatic changes was limited.

The conception of intrinsic reduction apparently grew out of the observation that water often seemed cooler than the air surrounding it (particularly in the summer). Another observation seemingly supporting this belief was that ice appeared colder than the air in which it was frozen. The conclusion drawn from these observations was that the temperature of water could be reduced below that of its container without the help of any external means, or, in other words, that the water cooled itself. The scholastics argued against the container cooling the water below its own degree from the Aristotelian standpoint that similars cannot act upon similars and that every action must take place as a result of a contrariety of qualities. The only way around this impasse was to assume intrinsic reduction.

Avicenna, so Paul of Venice tells us, was one of the first to offer an explanation as to why hot water seemed to reduce itself. He said that it did so by its form.[1] With this statement of Avicenna as authority the schoolmen set up explanations of

1 Paulus Venetus, *Expositio super libros de generatione et corruptione Aristotelis,* Venetiis, Bonetus Locatellus, 1498, f. 43r "Avicenna dicit primo sue Physice quod aqua calefacta per suam formam frigefacit se."

59

intrinsic reduction. These explanations were followed in turn
by indignant protests from the supporters of extrinsic reduction.

Before examining the various theories in any detail it can
be observed briefly that Albert of Saxony explains the phenomenon of reduction by assigning to the water "virtual
frigidity" which is aided in the reduction by the air and by
superior agents. Marsilius of Inghen falls back on the substantial form of the water as the cause, while Walter Burley
denies any explanation involving intrinsic reduction and finds
the cause exclusively in external factors. The Italians have a
tendency to compromise, as is seen in the theories of Paul of
Venice, Jacobus de Forlivio, and Giovanni Arcolani, who was
Marliani's opponent in discussing this question. Marliani, however, stands firmly for extrinsic reduction, bringing to the
problem his principle of active and resistive powers based on
the quantity of heat or cold.

In declaring that the reduction of hot water is accomplished
by its own form, Avicenna did not determine whether it is a
reduction by some formal or virtual frigidity.[2] However, Albert of Saxony is more explicit. He tells us that reduction takes
place as result of a dissimilarity, that is, by means of the virtual
frigidity which is dissimilar to the accidental calidity existing in the heated water.[3] By this statement Albert simply
means that water is naturally possessed of a powerful frigidity
or coldness which is able to overcome the heat temporarily in-
duced in the water by some heating agent. But the Parisian
schoolman does not deny that there are extrinsic factors which
aid the virtual frigidity in reduction. He points out that when

2 *Ibid., loc. cit.* "...non determinat (Avicenna) utrum ista reductio fiat
mediante frigiditate aliqua formali aut virtuali..." I have been unable to
read Avicenna's *Physica,* and as a result have had to rely on Paul of Venice's
summary of his position.

3 Albertus de Saxonia, *Questiones in...libros de generatione,* (with the
*Commentaria...D. Egidii Romani in libros de generatione et corruptione
Aristotelis,* Venetiis, Bonetus Locatellus, 1504, f. 141 v, c. 1) "...de aqua
vero calefacta, dico quod illa reductio ad frigiditatem pristinam sit ratione
dissimilitudinis. Nam illa fit per frigiditatem virtualem que est dissimilis
caliditati accidentali existenti in aqua calefacta."

ice is formed in the winter, the air, and the superior influences of the constellations take part in the freezing.[4]

Marsilius presents a solution somewhat similar to that of Albert, although he actually challenges reduction by virtual frigidity. In its place he substitutes the " substantial form " of the water as the cause of the reduction.[5] The substantial form is aided by superior agents in bringing about the cooling, once the heating agent has been removed.[6]

As an argument against Avicenna and his successors, Paul of Venice cites the Aristotelian belief that the agent and the patient must be contraries.[7] By this he means that water when it is heated has certain qualities, namely, wetness and hotness, and it cannot be expected to act upon itself. Only a contrary having the proper quality of coolness could cool the water.

The defenders of Avicenna, however, say that this statement of Aristotle is to be understood when a patient is in essential *potentia* for inducing form. When it is in accidental *potentia,* contrariety is not necessary. It will be able to move itself from accidental " potential " to accidental " act ", as Aristotle showed in the eighth book of the *Physics*.[8] Therefore, since heated water is accidentally potential to frigidity, it can reduce itself

4 *Ibid., loc. cit.,* " ... hoc non solum facit aer, sed ad hoc concurrent constellationes et influentie superiores."

5 Marsilius of Inghen, *Questiones in prefatos libros de generatione* (published with the edition of Romanus and Albertus of Saxony already cited), f. 105v, c. 2 " ... in aqua calefacta remoto calefaciente generatur frigiditas immediate a forma substantiali cum agentibus superioribus."

6 Paul of Venice (*op. cit., loc. cit.*) summarizes the position of Marsilius in the following way: " Ideo dicit Marsilius quod [reductio] sit per solam formam substantialem absque frigiditate formali aut virtuali, quoniam natura dedit forme substantiali instrumentum accidentale propter actionem transeuntem in materiam exteriorem. Sed huius actio non est in materia extranea, sed intranea cui ipsa forma est presens et indistans."

7 *Ibid., loc. cit.*

8 *Ibid., loc. cit.* In the *Physica* (VIII, 4, 255a-b) Aristotle, while trying to show how bodies return to their natural places, implies that a body having been accidentally moved from its proper place is accidentally potential for returning to that place, and as soon as the hindrance has been removed, it returns of its own accord to its natural place.

to frigidity when the heating agent has been removed. Cold water, however, which is essentially potential to calidity (i. e. hotness), can not reduce itself to calidity.[9]

This argument can be most successfully combated on the basis of experience. If it were true, water would always reduce itself to the greatest possible frigidity regardless of the season. Water would not be cooled any more in the winter than in the summer.[10] It can also be opposed on the ground that nothing is changed from essentially potential to act unless it is changed according to species. But hot and cold water are of the same species.[11]

Burley answers the objection that action must result from a contrariety in the most plausible manner by denying intrinsic reduction, and affirming in its stead extrinsic reduction. The air is not as hot as the water it surrounds. Therefore, there would be a contrariety and the water would be cooled. If some one should oppose this position by saying that the water is reduced to a degree of frigidity greater than that of the air, Burley would deny that this took place. The cooled water is

9 Paul summarizes this argument from Walter of Burley, who, as we shall see, discards it in favor of extrinsic reduction.. The argument is as follows (*op. cit., loc. cit.*) : " Respondet Burleus hic loco Avic. et omnium istorum dicens quod illud dictum est intelligendum quando patiens est in potentia essentiali ad formam inducendam. Quando autem patiens est in potentia accidentali non requiretur contrarietas. Immo idem poterit seipsum movere de potentia accidentali ad actum accidentalem, ut determinat Aristoteles octo physicorum. Quia igitur aqua calefacta est in potentia accidentali ad frigiditatem, ideo potest seipsam reducere ad frigiditatem, et quia frigida [aqua] non est in potentia accidentali, sed essentiali ad caliditatem, ideo non potest seipsam reducere ad caliditatem."

10 *Ibid., loc. cit.,* " Ista opinio non videtur esse vera, quoniam aqua semper se ipsam reduceret ad frigiditatem summam. Et per consequens non ad maiorem frigiditatem seipsam reduceret in tempore hyemali quam tempore estivo, quod est contra experientiam." This objection is not necessarily valid, since the supporters of intrinsic reduction do not deny the extrinsic action of warm air in the summer.

11 *Ibid.,* f. 43v, c. 1 " Preterea nihil movetur de potentia essentiali ad actum nisi transmutetur secundum speciem.... Sed aqua calida et aqua frigida sunt eiusdem speciei."

actually no colder than the air, but on account of its density it seems colder.[12]

However, in another place Burley, while still holding out for extrinsic reduction, admits that water can be cooled beyond the degree of the container. He explains this by saying that something less cold can cause greater cold in a subject having a greater aptitude *(aptitudo)* or inclination *(inclinatio)* for frigidity, just as a small fire can cause a larger fire in a subject more combustible.[13]

The obvious fallacy in this last argument presented by Burley is a confusion of intension and extension, or, as we would say it today, a confusion of temperature and quantity of heat. Yet Paul of Venice in discussing this second opinion does not challenge it on this ground, but rather unjustly with the inconvenience that was adduced against Avicenna. If reduction below the degree of the container is accepted, then we would expect water to be reduced to its greatest frigidity at all times, regardless of the season.[14]

Against Burley's first opinion, which explained reduction by external factors to the temperature of the container and no further, Paul likewise raises an objection. Such an opinion can not explain the fact that in the summer the air is hot and the reduced water is cold.

In view of the objections that Paul has raised to the opinions of Burley, we are not at all surprised to find that Paul's ex-

12 *Ibid.*, f. 43v, c. 1 " Secundus modus respondendi est Burlei dicentis hic quod reductio aque ad frigiditatem non fit ab intrinseco, sed ab extrinseco, scilicet a continente, ut puta ab aere et celo. Unde aer circumdans aquam calefactam non est ita calidus sicut aqua. Et immo cum habeat contrarietatem ad aquam, potest frigefacere illam. Et si arguitur in contrarium, quia reducitur ad maiorem frigiditatem quam sit frigiditas aeris, respondet Burleus quod non, quia aqua reducta ad frigiditatem non est frigidior aere, sed propter densitatem eius apparet frigidior cum virtus unita fortior sit seipsa dispersa. ... "

13 Walter Burley, *In octo volumina ... Aristotelis de physica auditu expositio ...*, Padua, 1476, f. 37r, c. 1. I have not been able to find in Burley's commentary on the *Physics* the passage quoted through Paul of Venice in note 12.

14 *Op. cit.*, f. 43v, c. 1.

planation of reduction represents a compromise including both objections. Reduction takes place both intrinsically and extrinsically, intrinsically *per accidens* and extrinsically *per se.* If the reduction did not take place intrinsically, we should be unable to understand why water is reduced to a greater degree of frigidity than that of the container. If it did not take place extrinsically, we should be at a loss to explain why the reduced water is colder in the winter than in the summer. The extrinsic container is a cooling agent *per se,* but is not necessary for the cooling of the water. Even if the water were placed in a vacuum, it would be reduced no less to frigidity, not indeed *per se,* but *per accidens* (no doubt in the manner that the defenders of Avicenna suggest).[15]

A more reasonable compromise than Paul's is that of Jacobus de Forlivio. Jacobus tells us that intrinsic reduction is impossible except as a result of a difformity of parts. If all the quantitative parts of man or any animal were uniformly hot, there would be no intrinsic alteration in them because there would be no contrariety, a *sine qua non* for any action.[16]

15 *Ibid.,* f. 43v, c. 2. " Concludo igitur quod ista reductio fit ab intrinseco et extrinseco; ab intrinseco quidem per accidens; et ab extrinseco per se. Nisi enim illa reductio fieret ab intrinseco non posset assignari causa propter quam reducitur ad maiorem frigiditatem quam sit frigiditas continentis. Nisi fieret ab extrinseco, non posset assignari causa propter quam est frigidior in hyeme quam in estate. Et licet continens extrinsecum sit infrigidans per se; illud tamen non est necessarium ad infrigidationem aque, quoniam si aqua calefacta poneretur in vacuo non minus reduceretur ad frigiditatem, non quidem per se, sed per accidens, sicut per accidens corrumperetur caliditas et generaretur frigiditas ... "

16 This statement follows after the question has been argued at considerable length (See the edition of Jacobus' *De intensione et remissione* published with the treatise of the same name by Walter Burley, Venetiis, 1496). The question arises when Jacobus adduces inconveniences against the first conclusion which declares that it is impossible for contraries to be in the same subject at the same time (f. 20v). An argument against this is the case of the intrinsic reduction of water where qualities corruptive of each other are together in the water (f. 23r, c. 2). This is challenged with the statement that reduction is intrinsic, against which it is immediately argued that the frigidity of the water is often more intense than that of the surrounding medium, " sicut apparet experimento de aqua generata in termis ex vapore

Up to this point we have noted the principal arguments used by those who discussed the question of reduction prior to its treatment by Arcolani and Marliani. The arguments have been largely philosophical, based on the Aristotelian conceptions of qualitative forms. With Marliani the discussion becomes more mathematical. He brings to this question as he did to that of reaction his theory of heat actions based upon calculated quantities of heat or cold. As before, he distinguishes between quantity of heat and intensity of heat (temperature). His conclusion, as we have previously noted, is that water must be cooled by external factors.

Marliani's opponent in this discussion, Johannes de Arculis of Verona (Giovanni Arcolani), a noted physician who taught at Bologna, Padua, and Ferrara, and who commented upon the *Canon* of Avicenna,[17] follows in the tradition of Avicenna by accepting intrinsic reduction. He varies, however, from the common opinions in the role he assigns to substantial form in this reduction.

This dispute of the two learned Italian physicians was engineered by a Doctor of Arts, Policletus ex Ferrariis of Mantua, who had, as he tells us, heard Marliani's opinions on the reduction of hot water while the latter was teaching medicine at Pavia. He sent these opinions to Giovanni Arcolani of Verona, who at that time was teaching medicine at Ferrara. The result of Policletus' efforts was this controversy, written down by Policletus in 1461 (See Chapter I, section F).[18]

aqueo ascendente et occurente parietibus termarum que pervenit ad frigiditatem valde sensibilem et tamen aer ibi existens et totum circumstans est sensibiliter valde calidum " (f. 23r, c. 2). After answering this argument of the condensed vapor in thermal baths, he notices that Marsilius urges reduction by substantial form, while others defend reduction by virtual frigidity (f. 25r, c. 2). But finally he concludes with the statement noted above in the text (f. 25v, c. 1).

17 See the *Biographisches Lexikon der hervorragenden Aertze aller Zeiten und Völker*, Vol. I, Berlin and Wien, 1929, p. 187 for the various editions of the commentaries on Avicenna.

18 The controversy was noted briefly by L. Thorndike, *A History of Magic and Experimental Science*, Vol. IV, New York, 1934, pp. 207-8. Thorndike

We are told, as the beginning of the dispute is described, that while the rector of the *studium* (of Pavia) was arguing in circles concerning the reduction of hot water to frigidity, he was answered by a certain doctor that hot water cooled intrinsically *ex se*. When Marliani, who had just approached, heard this response, he immediately offered arguments against it.[19] He first declares that he wishes to prove that heated water is not cooled *ex se* beyond the degree of frigidity existing in the air surrounding it. For the present he decides to omit the argument that he customarily uses to the effect that a large amount of water ought to be cooled more swiftly than a small, which is obviously against experiment.[20] He likewise omits the argument that if reduction proceeded intrinsically, all elements ought to be cooled to the degree of frigidity of the water.

Therefore, in the first phase of the argument Marliani limits himself to a mathematical example designed to prove that if intrinsic reduction is conceded, then action must proceed from a minor proportion, that is to say, the reduction would proceed, according to the Peripatetic law of the relation of active power to resistive power, from a proportion of lesser inequality (i. e. a proportion where the first term, the active power, is less than the second, the resistive power). We know, however, that an action can proceed only from a proportion of greater inequality *(maior proportio)* where the active power is greater than the

mentions here that Michael Savonarola, the Paduan doctor, apparently believed in the intrinsic reduction of heated water.

The introduction to this dispute informs us of the circumstances of its inauguration (Ms. San Marco, f. 12r, c. 1) : "Policletus ex ferrariis Mantuanus artium doctor insignis, hec infra scripta argumenta, que a Joanne Marliano mediolanensi medicinam papie legente audiverat, partim eo dicente, partimque disputante, hac serie scripsit clarissimo artium et medicine doctori interpretanti domino magistro Joanni Arnulfo de Arculis Veronensi, Ferrarie medicinam docenti, petens ab eo ad illa responsiones, quibus doctior fieret et tanti viri in hiis difficilibus que in illis tangunt argumentis sententiam haberet."

19 *Ms. cit.*, f. 12r, c. 1.

20 *Ibid.*, f. 12r, c. 1. " .... et volo pro nunc ommittere argumenta que adducere sum solitus de magna portione aque que cicius infrigidari deberet parva portione, ceteris paribus, quod est contra experimentum... "

resistive. Since the hypothetical case which is justifiable on the basis of intrinsic reduction cannot satisfy the correct proportions, then intrinsic reduction must be abandoned. In other words, Marliani's mathematical example is an indirect proof.

The hypothetical case set up by Marliani is somewhat involved and there is little point in describing it at length. It is sufficient to say that, like some of the cases cited in the preceding chapter and some of those which will be discussed later in this chapter, it illustrates Marliani's idea of heat actions based upon quantity of calidity and frigidity. It is concerned with a determined amount of water cooling intrinsically and its effect upon some pepper placed within it in a fixed quantity so that a proportion of one hundred to one is secured between the active power of the water for infrigidation and the resistive power of the pepper to infrigidation. Marliani's conclusion is that whatever action of the pepper takes place in the water, it will proceed from a proportion of lesser inequality. Hence he decides against intrinsic reduction below the degree of the container.

This mathematical case apparently baffled the doctor who had started the argument. He promised to return the next day with a solution, but at that time had not yet succeeded in finding one.[21]

In his first response Arcolani supports a position similar to that of Paul of Venice. In cases where the water heated above the temperature of the container is cooled, it is cooled both by the container and intrinsically. However, when the water is of the same degree of calidity as the container or less, then it is reduced intrinsically and not to the frigidity of the container, since a similar can not act upon a similar.[22] He elaborates upon

21 *Ibid.*, f. 12r, c. 2 "Ad hoc argumentum ille doctor dixit se in sequenti die responsurum. In die tamen immediate sequenti, nullam rectam viam habuit respondendi."

22 *Ibid.*, f. 12v, c. 1 " Tenendum est quod aqua calefacta calidior continente reducitur ad frigiditatem a frigiditate continentis et ab intrinseco. Si vero aqua calefacta sit in caliditate similis caliditati continentis aut frigidior, tenendum est quod reducitur ad frigiditatem ab intrinseco et non ad frigiditatem continentis, cum simile non agat in simile."

this position at some length, proceeding upon the assumption that the supporters of extrinsic reduction must explain how a container is able to cool water below the temperature of the container. In this sense, he is putting up straw men to knock down, for Marliani does not believe that such a reduction below the degree of the container takes place. In summing up the evidence against extrinsic reduction, Arcolani cites Galen and Avicenna as supporting his declaration that "something remitted can not intensify something intensified," a corollary of the belief that similars can not act upon similars, and simply means that we wouldn't expect something of little power to increase the power of something of great power.

From a general attack on the theory of extrinsic reduction, Arcolani turns to a specific criticism of the points raised by Marliani. He first assails the argument that Marliani had intended to omit, which claimed that if intrinsic reduction were assumed, then one would expect a large portion of water to cool more swiftly than a small. The greater power that is in a large quantity of water does not cool it more rapidly than a smaller amount, but actually more slowly because there are hot vapors in the water which are driven out as the water cools. It is more difficult for these vapors to escape from a large quantity of water than a small, both because they are more plentiful and because they are situated more deeply.[23]

As for Marliani's second contention, not elaborated upon, that in intrinsic reduction all elements would be cooled to the degree of frigidity of the water itself, Arcolani presents no answer because "no deduction of this consequence appears in Marliani's statements."[24]

---

23 *Ibid.*, f. 12v, c. 2 "...causa fovens caliditatem aque sunt vapores ad minima cum aqua calefacta permixti cuius signum, quia aqua calida semper effumat et quanto calidior, tanto magis effumat, et quanto magis infrigidatur minus effumat. Item qui vollunt infrigidare aquam ipsam agitant et commovent ut vapores exallent. Nunc autem quanto aqua est maioris quantitatis, tanto difficilius et tardius vapores exallant, tum quia plures, tum quia magis profundi."

24 *Ibid.*, f. 13r, c. 1.

The case that Marliani has posed for his main argument involving pepper and water is denied by his opponent. Arcolani
objects particularly to the series of minor proportions which
Marliani has derived from the relation of the water, cooling
intrinsically, to the pepper. The chief point in his argument is
that pepper can not be cooled except by the cooling of the water;
i. e., that the rate of cooling is a function of the water and
pepper together, irrespective of the subject.[25]

These statements have been included here from Arcolani's
criticism of Marliani's original hypothetical case because they
provide a spring board for what Marliani considers a very important refutation of Arcolani's conception of heat actions.
We saw in the chapter on reaction that Marliani had set up a
theoretical temperature scale in which temperature was considered as a balance between a certain number of degrees of
calidity and a number of degrees of frigidity, and as the
calidity approached the highest degree, correspondingly the coextensive frigidity approached the lowest degree. Water at any
temperature, measured under such a system, however, would be
at equilibrium. Therefore, in the case of hot water which has a
calidity of 7 degrees and a frigidity of one degree, the calidity
which is in equilibrium with the coextensive frigidity could
neither act upon nor suffer action from that frigidity. The only
way that the water could be cooled would be by some outside
agent which would be at a different temperature. Yet in the
case of intrinsic reduction, Arcolani clearly indicates that the
frigidity of the water, although meeting resistance from its
calidity, actually alters the temperature by itself.

Marliani will have a great deal to say against Arcolani's explanation of how frigidity can act on its coextensive calidity,
but he first attacks, point by point, Arcolani's general theory of
reduction, which he divides into three parts. The first part, in
which it was held that water heated above the degree of the

25 *Ibid.*, f. 13r, c. 2 "... non potest piper infrigidari nisi aqua infrigidetur
in qua positum est, et tantum precise resistat caliditas aque, ne aqua infrigidetur, quam ne piper infrigidetur."

container is cooled both extrinsically and intrinsically, evokes little discussion.[26] But the second, that the container can not cool the water below its own degree, is questioned. Marliani knows as well as Arcolani that similars can not act upon similars, but he also believes that water is not cooled below the degree of the container. Therefore, there is no reason to use this argument. Marliani confines himself at this point to discussing the problems that his opponent has used to prove his argument that similars do not act upon similars and that something " remitted " does not act upon something " intended ".

In the discussion of one such problem the Milanese physician makes clear his distinction between active power and intensity. An action of a frigidity of 3 on a calidity of 6 must not be considered out of the question (as Arcolani intimates), for, and the doctrine is attributed to the Calculator, something more remiss in intensity can be more powerful for acting or resisting. A body which has a frigidity of an intensity of only three degrees can be so large or so dense that it will cool a body which has a calidity of an intensity of 6 degrees, but which is small or rare.[27] We can conclude from this discussion that the power of acting or resisting is dependent on the quantity of heat or cold involved. This quantity of heat or cold in turn is dependent on the quantity (i. e. volume) and density of the body (which taken together represent the mass of the body) as well as the intensity of heat or cold.

The third part of Arcolani's theory of intrinsic reduction described the mechanics of this reduction. The Veronese phy-

26 *Ibid.*, f. 13v, c. 2.

27 *Ibid.*, f. 14r, c. 1. " ... Et cum inferebat quod si frigiditas *a* que est ut 3 agit in caliditatem *b* que est ut 6 a proportione minoris inequalitatis fieret motus ... negaretur consequentia ... Calculator noster evidenter probavit una qualitas erit altera remissior, que tamen erit eadem multo potentior in agendo et resistendo, stat ergo *a* habens caliditatem ut 5 et frigiditatem ut 3 esse tante quantitatis aut tantum depressum, et *b* habens caliditatem ut 6 et frigiditatem ut 2, esse ita rarum, aut tantum modice quantitatis quod multo maior frigiditatis ut 3 erit in *a* quam caliditatis ut 6 erit in *b*. Quare potentior erit frigiditas *a* ut 3 in agendo quam caliditas *b* intensa ut 6 in resistendo ".

sician had said that heated water which is cooled beyond the
frigidity of the container is cooled intrinsically because the
form of the water directs its frigidity to act on calidity which
is contrary to its form and is accidentally impressed upon it.[28]
First of all, Marliani accuses Arcolani of not adhering to the
common method of speaking of intrinsic reduction, which calls
for reduction by difformity of parts or by substantial form
either without any instrument or with virtual frigidity as its
instrument. In addition, Marliani points out that the frigidity
existing in the water is not contrary to the calidity coextensive
with it. Therefore, it does not act on it, since every action must
take place as a result of a contrariety. Or, as we have said before, the frigidity and coextensive calidity are in equilibrium
and this condition can not be altered except by another body
with a different degree of calidity and frigidity.

After directing another argument against Arcolani's theory
of the role of substantial form in reduction, Marliani turns
to a criticism of his theory of vapors. It is common knowledge
that the vapors are of the same nature as water. Therefore, since
they are more tenuous, i. e. their density is slight *(ad minima
divisi),* they will be cooled along with the water by their substantial form through the double instrumentation of the frigidity of the water and that of the container, assuming, of course,
that the vapors were hotter than both of them. Consequently,
the vapors will not remain hotter than the water,[29] and the
vapor theory can not be true.

Another argument brought against Arcolani's vapor theory
by Marliani assumes that heated water is frozen more swiftly
than cold water. If the vapor theory were true, such a phenomenon could not be explained. To support his contention that
heated water freezes more rapidly, Marliani first points to a

28 *Ibid.*, f. 14r, c. 2 " Tertio dicitur quod aqua calefacta cum infrigidatur
ultra gradum continentis infrigidatur hoc modo ab intrinseco, quia forma aque
derrigit (!) frigiditatem suam ad agendum in caliditatem sue forme contrariam
sibi accidentaliter impressam."

29 *Ibid.*, f. 14v, c. 1.

passage in Aristotle's *Meteorologica* affirming it.[30] However, the Milanese physician does not depend on Aristotle's statement alone. He claims that not only has he often tested its truth during a very cold winter, but that anyone may do so. You take four ounces of boiling water and four ounces of non-heated water and place them in similar containers. Then the containers are exposed to the air on a cold winter's morning at the same time. The result is that the boiling water will freeze the faster.[31]

This experiment of Marliani's is not as mysterious as it sounds. What happened, of course, was that the four ounces of boiling water evaporated rapidly, and, in doing so, the mass of the water was decreased appreciably. The cold water, on the other hand, evaporated slowly. Therefore the boiling water of decreased mass reached the freezing point first. This experiment will not work if the water is not at boiling temperature.

Marliani, it is true, was unable to find the cause of this phenomenon. If he had used what he formerly described as common knowledge, that " vapors are of the same nature as water ", and had been able to weigh the ice in both containers after the completion of the experiment, he would have been able to arrive at the correct solution, since he knew that a smaller quantity of water would freeze more rapidly than a large. At any rate, Marliani rejects the common explanation that heated water freezes more quickly because it is rarer. The solutions of the " Ancients " recited by Avicenna, as well as

---

30 *Ibid.*, f. 14v, c. 1. This statement is one of the arguments used by Aristotle and the medieval schoolmen to prove antiperistasis (See Chapter IV *infra*, note 56 and text).

31 *Ibid.*, f. 14v, c. 1 " Hoc idem in yeme multum frigida sepius sum expertus et quilibet experiri poterit. Nam si accipiantur quatuor unctie (*sic*) aut circa ex aqua putei vel fontis et non ultra calefiat, et in aliquo vase, autem multum frigido, ut in aurora diei hyemis multum frigide exponitur, et alia aqua eiusdem putei vel fontis, bulliat vel multum ferveat ex qua calida quatuor uncie etiam reponantur in vase omnino simili priori et simul exponantur aeri, et sit in omnibus aliis paritas, multo cicius congelabitur aqua fervens ... quam aqua non aliter calefacta."

that of the Moslem philosopher himself, also seem inadequate to the Pavian doctor.[32] As a result he offers no explanation.

After this rather interesting discussion of Arcolani's vapor theory and its consequences, Marliani explains how he would derive the conclusion that all elements would be cooled to the degree of the water if the water is cooled intrinsically beyond the degree of the air surrounding it. The water cooled below the temperature of the air would have to be cooled in spite of the air. We would expect, therefore, that after cooling to this lower temperature, the water would be able to cool the air immediately surrounding it, which in turn would cool more air adjacent to it, etc. It is known that air cools fire. As a final result, all elements would be cooled by the water, which is of course absurd.[33]

From this dialectical argument Marliani turns to a further defense of his original hypothetical case involving pepper and water. He is concerned especially with refuting the arguments of Arcolani which allowed action between a frigidity and its coextensive calidity. This refutation takes the form of several mathematical problems. These problems furnish us further evidence of Marliani's theory of heat actions.

We are told that there are two fundamental suppositions upon which Marliani's problems of heat action are based. The first is that latitudes of prime contrary qualities are interconnected *(complecte)*. The second is that in a uniform medium the more powerful or strong a uniform spherical agent is, the greater is the distance at which it can act.[34]

By the first supposition he means that, according to the system of coexistent contrary qualities already outlined in the

32 *Ibid.*, f. 14v, c. 2 "Neque dicta antiquorum ab d. Avicenna recitata secunda primi, neque dicta Avicenne ibi hanc difficultatem plene salvabunt.

33 *Ibid.*, ff. 14v, c. 2-15r, c. 1.

34 *Ibid.*, f. 15v, c. 1 "...presuppono quod latitudines primorum contrariorum sint complecte. Secundo presuppono quod quanto agens uniforme spericum fuerit fortius et potentius in medio uniformi, et ceteris paribus, tanto agat ad longiorem distantiam..."

preceding chapter,[35] the alteration in intensity (latitude) of the
one quality brings about a corresponding alteration in the opposite direction of the intensity of its coexistent contrary
quality. The system as applied to temperature envisions a set
number of degrees running from the *summus* (highest possible) degree of frigidity, which is at the same time the zero
degree of the coexistent calidity, to the zero degree of frigidity,
which is also the *summus* degree of the calidity. We have seen
that the scale of degrees employed by Marliani was ordinarily
eight, so that the relation of heat intensity (C) to coexistent
cold intensity (F) could be expressed by $F = 8 - C$.

Uniting what the second supposition tells us to the information that can be gleaned from the various problems cited here
by Marliani, we discover that a heat action is dependent on the
following factors: the absolute powers of calidity and frigidity
(which are dependent on the quantities of heat and cold), the
shape of the agent (and, for that matter, of the patient also),
the medium of the action, and finally the distance of the agent
from the patient.[36]

In his second set of responses Arcolani suddenly starts to
defend his belief in action between contraries rather than

35 See Chapter II, note 4, and text.

36 The first problem shows the power as a function of the quantity of
cold (*multitudo frigiditatis*) or heat, which is itself clearly shown to be an
extension of the intensity of cold or heat. It also illustrates action through
a certain distance in a given medium (air). *Ibid.*, f. 15v, c. 1 "Quibus ita
premissis pono quod *a* aqua magne quantitatis habeat per totam caliditatem
ut unum et frigiditatem ut 7 et sit hec aqua tante quantitatis ut habeat tantam
multitudinem frigiditatis ut sit illa frigiditas potentie ut 70 et eius caliditas
solum potentie ut 10.... Et ponamus quod aer sit similis aque vel parum
calidior, et agat gratia argumenti (aque caliditas) potentia ut 70 in illi (!)
aere per 70 pedalia..." In another problem involving an iron spherical agent,
Marliani brings in the density of the medium as a factor in determining the
distance through which a given power can act. *Ibid.*, f. 15v, c. 2 "... pono
quod *a* sit unum ferrum spericum magne quantitatis, habens caliditatem ut
7 et frigiditatem ut unum. Et sit in ipso tanta multitudo caliditatis ut sit
potentie ut 70 et erit eius frigiditas potentie ut 10 vel minoris quam ut 10.
Et pono ultra quod in *b* medio uniformi multum raro potentie ut 1000 [la] agat
caliditas potentie ut 70 species suas per distantiam 70 pedum..."

similars. He points to his views as outlined in a work entitled *De compositione medicinarum*. It does not seem likely, however, that Marliani actually thought that the Veronese physician believed in action between similars, but only that certain logical deductions from intrinsic reduction would lead to such a belief. At any rate, after defending his views by reference to the Calculator, Arcolani accuses Marliani of misrepresenting his views and urges him to consult the above-mentioned work.[37]

Arcolani's responses are here much shorter than the previous ones. He first repeats his theory of how intrinsic reduction takes place. The water is reduced to the frigidity natural to it by the small amount of frigidity coextensive with the calidity which has been acidentally impressed on it. This frigidity is directed by the form itself of the water to corrupt the calidity made disproportionate to it.[38] Marliani's criticism that this is not the common method of explaining reduction is granted, but Arcolani justifies his method by saying that it seems to him to fit the circumstances better than the other methods. Marliani's second objection, that the frigidity existing in the water is not contrary to the calidity coextensive with it, he denies, claiming that every calidity is contrary to every frigidity.

We might wonder, as Marliani had, how, in view of this assertion that calidity is contrary to its coextensive frigidity, water with a calidity of 6 and a coextensive frigidity of 2 could ever be cooled. Arcolani replies that the union of the form of the water with the frigidity of 2 is more powerful than the calidity of 6.[39]

After this rather feeble answer, Arcolani turns to a defense of his vapor theory, which had proposed that the greater quantity and deeper situation of vapors in a larger quantity of water were the causes of its cooling more slowly than a smaller quantity. In criticism of this theory it was suggested

---

37 *Ibid.*, f. 17r, c. 2.

38 *Ibid.*, f. 17r, c. 2.

39 *Ibid.*, f. 17v, c. 1 "... respondeo quod aggregatum ex forma aque et frigiditate ut duo est potentius quam caliditas eiusdem aque ut 6 ... "

that the shapes of the vessels containing the different quantities of water could be so regulated that their depths and widths would be equal and their lengths unequal. Vessels shaped in this way would control the escape of hot vapors and thereby eliminate them as a factor in the cooling. Therefore, assuming that the hot water in each vessel started cooling from the same temperature through the intrinsic factor, the rate of cooling should be the same in each vessel. This, however, is contradicted by experiment. The larger quantity of water still cools more slowly. But, according to Arcolani, this experiment does not endanger the vapor theory. The larger quantity of water is cooled more slowly because it heats its container (the actual container plus the air surrounding it) more than does the smaller amount. The heated container then retards the intrinsic reduction.[40] In his last set of responses Marliani rejects such an explanation on the grounds that the greater power of the larger quantity of water will encounter proportionally more resistance both from the air and the sides of the container than will the smaller quantity.[41]

Of the remaining answers which Arcolani gives to Marliani's arguments, we shall note only one. Marliani, as we have seen, had advanced as an objection to the vapor theory an experiment which apparently proved that boiling water froze more rapidly than non-heated. He had professed his ignorance of the cause of this phenomenon. Arcolani, on the other hand, readily offers an explanation. " *a* (the heated water) is frozen with less frigidity than *b* (the non-heated). It is not necessary that congelation or density be acquired in these two quantities of water in the same proportion as frigidity is acquired . . . for oil is frozen with less frigidity than water, and one oil with less than another, as daily experience shows us." [42] Marliani, however, in

40 *Ibid.*, f. 17v, c. 2.

41 *Ibid.*, f. 20v, c. 1 " Nam si potentia multe aque erit maior, ita potentia aeris applicati multe aque erit maior, quia proportionaliter plus de aere obicietur, et etiam de parietibus vasis aque multe quam parve ... "

42 *Ibid.*, ff. 17v-18r " Et ad rationem dicitur quod *a* cum minori frigiditate

his final set of responses points out that the heated and non-heated water are of the same species and complexion with no occult qualities, while oil and water are not. In the latter case we would naturally expect different freezing rates.[43]

Arcolani makes no attempt here to respond to the various mathematical problems which Marliani had cited. In fact, he concludes by praising Marliani's ability in calculations and geometry.[44]

Marliani's third and final group of arguments is a rather long summary of those which he had previously advanced. We have already had the occasion to examine a number of them, so that we can safely omit the greater part.

We are told in the first place that Jacobus de Forlivio and "many others" offer sufficient proof that calidity and frigidity existing together coextensively are not contraries (i. e. would not act on each other), but only of the same species as contraries (i. e. would take part in an action with external contraries).[45]

In discussing further the consequences of intrinsic reduction, Marliani claims that certain absurd problems could be proved. One of these shows that if intrinsic reduction were accepted, then you could place six ounces of water which is quite cold in an oven which is hot enough for cooking bread, and yet these six ounces would be cooled further *ex se*.[46]

Arcolani had many times tried to place Marliani in the position of accepting the contention that the container cools the water beyond the container's own degree, in order that he could accuse him of accepting the action of similars on simi-

congellabitur quam *b*. Neque est necessum, quod equaliter et eque proportionaliter in hiis duabus aquis acquiratur congellatio aut dempsatio, sicut acquiratur frigiditas ... oleum enim cum minori frigiditate congellatur quam aqua, et unum oleum quam aliud, ut quottidiana declarat experientia."

43 *Ibid.*, f. 21r, c. 2.

44 *Ibid.*, f. 18r, c. 2 "Sed in hiis cognovi te doctum non solum in calculationibus sed etiam in geometria."

45 *Ibid.*, f. 19v, c. 1. See Chapter II *supra*, note 4.

46 *Ibid.*, f. 20, cs. 1-2.

lars. But, like Walter Burley, Marliani refuses to accept this contention. He denies categorically that he has ever said anything from which it would follow.[47]

With these three arguments drawn from the final responses of Marliani, we conclude the discussion of reduction by the schoolmen of the fourteenth and fifteenth centuries. It should be noticed in Marliani's favor that he accepted without question the external cooling of hot water, rejecting the contention that water could be cooled by the container below the temperature of the container. He elaborates further in this disputation upon the fundamental factors involved in heat actions. In doing so, he distinguishes between the intensity of heat or cold (temperature) and the extension of that intensity (quantity of heat or cold). Likewise, it is shown that the mass of a body is important in " the quantity of heat or cold " which it possesses.

47 *Ibid.*, f. 20v, c. 2 " Nec aperte dixi aquam ultra frigiditatem continentis a continente infrigidari neque aliquid ex quo sequatur."

CHAPTER IV

BODY HEAT AND ANTIPERISTASIS
FINAL SPECULATIONS ON HEAT

In his last work, *Questio de caliditate corporum humanorum,*
written in 1472, published in 1474 and 1501 (see Chapter I,
section J), Marliani extends his speculations on heat to the
physiological realm of the human body, making a study of body
heat and temperature. In this work he attempts to bring order
in the controversy centering around the question " whether the
human body is hotter in the winter than it was in the summer
preceding." At the same time he discusses the kindred physical
question, how it is possible in ordinary inanimate bodies for
antiperistasis to take place. Antiperistasis is defined as the supposed sudden increase of the intensity of a quality as a result
of being surrounded by its contrary quality, for instance, the
sudden heating of a warm body when surrounded by a cold.

In this work as in the preceding ones, Marliani carefully
distinguishes between intensity and quantity of heat. Therefore
there are two questions to be determined. First, is the body
intensively hotter (i. e. has a higher temperature) in the
winter than in the summer? Second, is the body naturally hotter
in the winter; or to put the question in another way, is there
a greater quantity or production of heat (i. e. a higher
metabolic rate, in modern terminology) in the body in the
winter than in the summer?

His treatment is based most immediately on those of three
Italian physicians preceding him by a generation or more. The
first of these is Jacobus de Forlivio (d. 1414), the physician
to whom we have already frequently referred in the preceding
chapters;[1] the second, Johannes Sermoneta (de Sermoneto, de

[1] See Chapter II, note 22 and its discussion, and Chapters II, III, V,
passim.

Sulmoneta), who was at the University of Padua in 1411 and taught at Bologna in 1431;[2] and last, Hugo Senensis (Ugo de Benziis), who variously taught at Pavia (1396-1399), Piacenza (1399), Pavia (1404), Bologna (1404-1405, 1410-1413), Florence (1421-1422), Bologna again (1423), Padua again (1423-1424), Bologna again (1424-1425), Pavia once more (1425-1429), and finally Padua for the last time (1430-1448?).[3]

Aside from these men to whose conclusions Marliani attempted to conform his own, the Milanese physician makes constant use also of the standard medical authors: Hippocrates, Galen, Avicenna, and Peter of Abano (cited as " Conciliator "). Additional references have been made to a work of Gentile da Foligno De resistentiis,[4] to opinions of Jean Buridan (mentioned as " Bridanus "),[5] to two translations of the Problemata doubtfully attributed to Aristotle, the translatio antiqua of Peter of Abano [6] and the translatio nova, which was probably

2 John is spoken of in the Paduan records as a Doctor of Arts and student in medicine. See C. Zonta, Acta graduum academicorum gymnasii Patavini ab anno MCCCCVI ad annum MCCCCL, Patavii, 1922, pp. 39, 58, 71. The explicit of John's Questiones super librum aphorismorum, etc., Venetiis, Bonetus Locatellus for Octavianus Scotus, 1498, remarks that these questions were disputed at Bologna in 1430. John's name, however, appears only in the roll of professors for the year 1431-1432. See U. Dallari, I rotuli lettori . . . Studio Bolognese dal 1384 al 1799, IV, Bologna, 1924, p. 61.

3 This itinerary has been traced from the following references: Codice Diplomatico dell'Università di Pavia, I, Pavia, 1905, pp. 336, 395, 425 et al. in index; II, Pavia, 1913, pp. 73, 223, 246, et al.; C. Zonta, op. cit., pp. 148, 190 et al.; U. Dallari, op. cit., IV, pp. 25, 48, et al.; A. Gherardi, Statuti della Università e Studio Fiorentino . . . Seguiti da un'appendice di documenti dal MCCCXX al MCCCCLXXII, Firenze, 1881, pp. 396-400, 402. The Piacenza reference has been cited among the Pavian documents.

4 Questio de caliditate corporum humanorum, Milan, 1474, f. 27v, c. 2. There are manuscripts at Erfurt (Amplonian F 251, ff. 196v-201) and Wiesbaden (no. 60, ff. 24-30r) of Gentile's De resistentia membrorum.

5 De caliditate, f. 20v, c. 1; f. 40v, c. 1.

6 Ibid., f. 30r, c. 2; f. 30v, c. 2; et passim.

that of Theodore Gaza,[7] and to the *Problemata* falsely ascribed to Alexander of Aphrodisias.[8]

Marliani expresses his particular debt not to these past authors but to his colleagues of the University of Pavia, and especially to the ducal physicians Ambrosius Griffus and Lazarus Thedaldus of Piacenza, whose opinions and judgments were always at his disposal.[9]

The discussion of body heat, and particularly the question as to whether it was greater in the winter than in the summer in ordinary healthy individuals, arose from a statement made by Hippocrates in the *Aphorisms* to the effect that the stomach and the bowels are naturally hottest in the winter and spring and that because of this increased innate heat, more food is required in these seasons.[10]

Jacobus de Forlivio in commenting upon this aphorism admits first that there are two approaches to this problem, a physical one involving a discussion of antiperistasis, or how a quality can be increased intensively in a body by contact with its contrary quality, and a medical approach paying strict attention to the human body, its component members and operations.[11] This double treatment was taken up by Marliani to

7 *Ibid.*, f. 30v, c. 1. There seems to be little doubt that the new translation is that of Theodore Gaza; for I have found all of Marliani's references in that translation. It was customarily called the "new translation" in the early editions. I do not exclude the possibility, however, that the "new translation" might be that of George of Trebizond, made somewhat earlier. I have been unable to read this translation for Marliani's references.

8 *Ibid.*, f. 30r, c. 2; *et passim*. The translation that Marliani used of this work was possibly that of his student, George Valla, who explains in the preface to his translation, addressed to Marliani, that it was as a result of Marliani's urging and under his auspices that the translation was made. *Problemata e graeco in latinum traducta per Georgium Vallam*, Venetiis, Antonius de Strata, 1488, f. 2r.

9 Gio. Mari., *De calid.*, f. 1r, c. 1.

10 Hippocrates, *Aphorisms*, I, xv, Loeb Classical Library edition, London, 1931, pp. 104, 106; 105, 107.

11 Jacobus de Forlivio, *Questiones super aphorismos Hippocratis cum supplemento questionum Marsilii de Sancta Sophia*, Venetiis, Bonetus Locatellus, 1495, f. 10v, c. 1 (questio 28).

form the general plan of his work. At this point we are interested only in Jacobus' second or medical approach.

As a first premise, Jacobus makes a necessary distinction between " intensively hotter " and " naturally hotter ". A body is said to be intensively hotter than another when subjectively it has a more intense heat than another. A body on the other hand is said to be naturally hotter when it has in itself proportional to its mass more natural heat, i. e. more *spiritus,* which is the instrument of the soul in its operations.[12]

From this premise we deduce that body heat, i. e. the further production or quantity of body heat is dependent on the quantity of the spirits. This *spiritus* is a favorite device of medieval physiology used to explain how the soul can be the seat and director of certain body functions. The soul wills the function; the *spiritus* carries it out.[13]

In the second premise Jacobus indicates that the *spiritus* and the blood with which it is associated are augmented in the winter.[14] If this passage means, as it seems to, that with the increase in the quantity of heat brought about by the increase in *spiritus,* an increase in the *volume* of the blood takes place at the same time in the winter, then of course this part of the theory is completely wrong. Actually, high temperatures cause an increase in the volume of the blood, since the blood is diluted by fluid drawn into the circulation from the tissues. At low temperatures the volume of blood is less, while the blood is more concentrated, and the percentage of blood solids is in-

12 *Ibid.,* f. 11r, c. 1, " Corpus enim alio (intensive) calidius ceteris paribus est ipsum in se habere subjective calorem intensiorem quam aliud sufficientem ipsum denominare calidum. Corpus vere esse calidius alio secundum naturam est ipsum plus in se proportionaliter respectu sue molis habere de calido naturali, scilicet de spiritu qui est instrumentum anime in suis operationibus."

13 I point here to an interesting passage from Professor Thorndike's *History of Magic and Experimental Science,* Vol. I, New York, 1923, pp. 657-660, which describes Costa ben Luca's theory of the *spiritus,* as well as those of other medieval writers.

14 Jacobus de Forlivio, *op. cit.,* f. 11r, c. 1, "... tempore hyemis in corpore robusto hyeme suam servante naturam et corpore debito utente regimine augmentatur spiritus et sanguis respectu eius qui fuit in estate et autumno."

creased. The change in blood volume is a very important factor in the control of body temperature.

The first conclusion of Jacobus tells us that bodies are intensively hotter in the summer than in the winter.[15] This conclusion, stated as it is categorically, has little value. The physicians following Jacobus, including Marliani, were to alter and amend this statement so that it referred only to certain exterior members.

The second conclusion is substantially that of Hippocrates, namely, that healthy bodies are naturally hotter in the winter than in the summer, and refers to the greater productivity or quantity of heat that exists in the body in the winter.[16] This is substantially true when we realize that due to the greater heat loss in the winter, the metabolic rate (heat productivity) rises. In fact, the rate begins to rise when the air temperature falls to around 68° F.

His third conclusion notes that there are some bodies that are both naturally and intensively hotter in the summer than in the winter. He cites in confirmation bodies that are so weak in heat or so sick that external heat can supply the needed heat.[17] This conclusion can be supported, for it is possible that if the temperature in the summer is sufficiently high, then the temperature of the tissues will be raised, the velocity of the chemical reactions will be increased, and as a result the heat production also increases.[18] It would likewise be possible to envisage the rise of the body temperature in this circumstance, since by

15 *Ibid.*, f. 11r, c. 2. " Corpora quecumque tempore estatis sunt intensive calidiora quam tempore hyemis."

16 *Ibid.*, f. 11r, c. 2. "... corpora robusta hyeme servantia suam naturam et utentia regimine debito [ut] supra sunt calidiora secundum naturam tempore hyemis quam estatis."

17 *Ibid.*, f. 11r, c. 2. " Tertio conclusio est: licet ita sit aliqua tamen sunt corpora in quibus in estate est calor maior tam intensive quam secundum naturam quam in hieme. Patet de corporibus debilis caloris et decrepitis quorum calor indiget confortatione ab extrinseco.

18 C. H. Best and N. B. Taylor, *The Physiological Basis of Medical Practice*, Baltimore, 1937, pp. 992, 998.

the rise in the environmental temperature the heat loss is reduced, while the heat production is increased.

Like Jacobus, the second of the Italian physicians, Hugh of Siena, distinguishes between intensively hot and naturally hot in commenting upon the aphorism in question.[19]

Hugh adds an interesting statement connecting the pores with natural heat. He declares that bodies have more natural heat in the winter because the pores are closed by the cold air, and the *spiritus* are prohibited from escaping, and thus they concentrate within.[20] In spite of the erroneous conception of *spiritus,* etc. this is an obvious recognition of the function of the pores in cutting down heat loss in the winter.

Hugh's general conclusions are those of Jacobus. He believes that intensively a body is apt to be colder in the winter than in the summer (i. e. like Jacobus he places environment as the determinant in body temperature), while naturally a body will be warmer in the winter than in the summer.[21]

On turning to Johannes Sermoneta's commentary on the aphorism in question, we find that of the three physicians he alone held out firmly for a constant body temperature in winter and summer. Like the others he adopted as a basis the common distinction between " intensively hot " and " naturally hot ".[22] John declares that the body temperature, a healthy body assumed, of course, is the same, regardless of the age of the individual, the season of the year, whether he is hungry or thirsty, or whether the body is exercising or at rest.[23]

19 Hugo (Bencius) Senensis, *Super aphorismos Hippocratis cum commento Galeni,* Ferrara, Lorenzo Rossi with Andreas de Brassis, 1493 (no pagination) sign. mark " e 3 ". " Notandum secundo quod ventres in hiis temporibus esse calidissimos potest intelligi duobus modis : uno modo intensive, et hoc non est verum ut apparebit in una questione, alio modo quia plus eis assistat de spiritu consanguine et corporibus calidis naturaliter iuvantibus ad operationes naturales et hoc modo intellexit Ypocrates."

20 *Ibid., loc. cit.*

21 *Ibid.,* signature mark " e 4 ", c. 2.

22 Johannes Sermoneta, *Questiones super librum aphorismorum,* etc., Venetiis, Bonetus Locatellus, 1498, q. xvii, f. 17r, c. 2.

23 *Ibid.,* f. 17v, c. 2. " ... a principio nativitatis usque ad finem vite servata

This statement for all practical purposes is true. We know, of course, that there is a variation of one degree between the temperature taken at the mouth and the rectal temperature, and likewise about the same variation in the opposite direction in mouth and axillary temperatures. There is also a temperature difference of about one degree between the maximum temperature in the afternoon and the minimum temperature in the morning, while strenuous exercise might cause a temporory rise of from one to four degrees (F°), etc. But in spite of these variations, we must conclude that John's statement is a recognition that there is some mechanism (or, as actually is the case, mechanisms) that tends to keep the body temperature even. While today we center heat regulation in the hypothalmic and other brain regions, John explained the constancy of temperature by means of unalterable, fundamental complexions *(complexiones)* acquired by generation.[24] Marliani was to make this conception of *complexio* the basic consideration in his treatment of body temperature.

Of considerable interest also is John's second conclusion in which he claims that the heating takes place at a much faster rate in the winter than in the summer, a recognition of what we call increased metabolic rate.[25]

His third and last conclusion repeats the second premise of Jacobus. Healthy bodies have a greater abundance of blood and *spiritus* in the winter than in the summer. He cites as a supporting fact that such bodies are fleshier in the winter.[26]

simili denominatione sanitatis in eodem corpore, tam in hyeme quam in estate, tam in fame quam in siti, tam in exercitio quam quiete, naturalis calor est intensive equalis."

24 *Ibid.*, f. 17v, c. 2. "(conclusio) probatur supponendo ad presens complexionem radicalem a generatione aquisitam non posse permutari..."

25 *Ibid.*, f. 17v, c. 2. "...Ergo crescit velocitas motus et per consequens calefactio erit multo velocior in hyeme quam in estate. Forte dicitur quod licet potentia crescat in hyeme magis quam in estate."

26 *Ibid.*, f. 17v, c. 2. "Tertia conclusio: abundantiora sunt in sanguine et spiritu corpora robusta tempore hyemis quam tempore estatis. Patet quia talia corpora sunt carnosiora tempore hyemis quam tempore estatis..."

Marliani's exposition of the question is considerably longer than those of his predecessors. According to the scholastic form, he adduces first common arguments against the question that bodies are either hotter in the winter than in the summer, or are hotter in the summer than the winter. The negative argument follows the line, first of proving that they are equally hot in summer and winter, second of proving against the first part of the question that they are hotter in the summer than the winter, and finally proving against the second part of the argument that bodies are hotter in the winter. Having disposed of the common negative arguments, he begins in the second part of the work (f. 7ff.) a preliminary treatment of the question with some conclusions that he was accustomed to reach in disputation. He explains that he was moved to compose the present work because a certain famous doctor brought out a treatise on the subject which argued against his conclusions. So, after outlining his preliminary conclusions in the second part, Marliani proceeds in the third (f. 8v. ff.) to a criticism of the discussion of this unidentified doctor, and finally in the fourth (f. 27r. ff.) to some further conclusions of his own. The fifth part (f. 39r. ff.) is a discussion of antiperistasis; while the treatise is concluded by a brief criticism of the original negative arguments (ff. 56-61).

Let us concentrate our attention on the second, fourth, and fifth parts, as they most clearly reveal Marliani's opinions on body heat and antiperistasis, with only slight reference to the other sections.

In the first part he notes one line of argument that is commonly used to prove that human bodies are equally hot in winter and summer. The body heat is constant because the spirits, which are the hot matter in our body, are unchanged in heat.[27]

---

27 G. Marliani, *De caliditate corporum humanorum*, f. 1v, c. 2. "Item tertio spiritus non sunt calidiores in estate quam in hieme neque econtra, ergo neque corpora humana sunt calidiora in estate quam in hieme nec in hieme quam in estate..."

The arguments in the first section are advanced without the careful premises laid down when Marliani discusses his own theory in the second and fourth sections. The common distinction between intensive and natural heat rarely appears in these negative arguments. Yet this distinction advanced by the earlier physicians is cited by Marliani as his first and basic premise in the second part. He tells us that there are two ways for one body to be hotter than another, absolutely and naturally. A body is said to be hotter than another absolutely when it has a more intense degree of heat, and he goes on to explain how this is true in cases of the existence of coextensive heat and cold in a body. The body which is hotter according to nature, on the other hand, is that which has a greater quantity of blood, natural heat, and *spiritus*.[28]

The second premise notes that in internal regions of the body there is a greater quantity of natural heat or of *spiritus* and blood in the winter than in the summer.[29] The third claims that certain healthy human bodies are not very rare in pores or in (skin) texture; while certain weak bodies are very porous or of broad pores or rare in texture. There are others between these two types.[30]

Connecting the pore theory with distribution of blood and natural heat, Marliani believes in his first conclusion that because of the broadness of their pores certain weak bodies will be hotter naturally in the summer than they were in the winter.[31] These weak bodies are those which are naturally cold. The cold air in the winter affects them considerably, and they will be unable to increase the production of heat (i. e. *spiritus*) at that time. Hence in the summer when they do not find it necessary to fight environment, heat production is easier.

28 *Ibid.*, f. 7r-v.

29 *Ibid.*, f. 7v, c. 2. " ... in interioribus maiorem quantitatem caloris naturalis sive spirituum et sanguis tempore hyemis reperiri quam tempore estatis fuerit ... "

30 *Ibid.*, f. 7v, c. 2.

31 *Ibid.*, ff. 7v-8r. "Aliqua corpora humana debilia magis calida sunt aut erunt naturaliter in estate quam fuerint in hieme proxime precedenti ... "

A second conclusion notes that there are certain sick or weak people, however, those who are spoken of as choleric people, who are less hot naturally in the summer than in the winter.[32]

The third conclusion is that of Hippocrates as presented by the medieval physicians, normal healthy bodies are naturally hotter in the winter than in the summer.[33]

These three preliminary conclusions treat of the natural heat of bodies. The fourth remarks that although not every part of the human body is intensively hotter in the winter than in the summer, Marliani feels it necessary to hold, against the opinion of many, that some internal parts are intensively hotter in the winter than they were in the summer.[34]

This conclusion in itself does not state that those members or organs which are not hotter in the winter than the summer are of constant temperature. No, from all we know they might be considered hotter in the summer than in the winter, as Jacobus wished in his first conclusion. We must wait until our examination of the fourth part of the treatise for this information.

All of these preliminary premises and conclusions, Marliani says, allowed him to hold to the common way *(via communis)* from which he never strayed unless sufficient reasons *(rationes)* or authorities *(auctoritates)* strongly compelled him to do so.[35]

At this point he declares that neither the conclusions of Jacobus, nor those of Hugo and John Sermoneta, nor, in fact, of any other ancient expositor, go against *(repugnantes)* his own conclusions.[36]

While discussing in the third section the premises of the theory he attributes to a famous doctor, Marliani remarks that the second premise is based on the theory that the power

---

32 *Ibid.*, f. 8r, c. 1.

33 *Ibid.*, f. 8r, c. 1. "...corpora humana robusta...sunt naturaliter calidiora in hieme quam fuerint in estate proxime precedenti..."

34 *Ibid.*, f. 8r, c. 2.

35 *Ibid.*, f. 8r, c. 2.

36 *Ibid.*, f. 8v, c. 1.

of something *(potentia rei)* is dependent on the amount of its form *(multitudo forme)* and that he himself has long been accustomed to say just that. As confirmation he points to statements in his *Liber diversarum conclusionum.*[37]

After analysing step by step the famous doctor's theory, Marliani proceeds to add some additional premises and conclusions of his own.[38] The first of these added premises poses that there exists in the members of a healthy human body, in addition to the first qualities by which they act and resist, a *complexio* or second quality, which is neither formally nor really any of the first qualities. By means of this *complexio* the members perform or resist diverse operations at diverse times.[39] This premise is supported by *experimenta* and references to Avicenna and Gentile da Foligno.

We learn more concerning the complexions of members from the succeeding premises. The second declares that although there is great diversity among complexions, any particular complexion is of a definite power for acting and resisting, or for conserving its subject in the disposition (or condition) natural to it.[40] The third admits that while the complexion is generated from a varying action of prime qualities *(ex varia actione primarum qualitatum)* or a varying mixture of miscibles, not always does a variation of prime qualities in a subject produce a proportional variation of the complexion in that same subject. Again Avicenna is his authority.[41] We can see that this premise

37 *Ibid.*, f. 9r, c. 2.
38 I might add that this doctor of whom Marliani speaks, and whom I have not identified, probably raised this question in a commentary on the *Aphorisms*. Marliani identifies it as the ninth question without any further illumination. *Ibid.*, f. 8v, c. 1.
39 *Ibid.*, f. 27r, c. 2. "... in membris corporis humani sani ultra primas qualitates in eis existentes per quas agunt et aliquando resistunt etiam ponitur complexio que est una qualitas secunda et non formaliter neque realiter alique primarum qualitatum, per quam membra agent diversas operationes in diversis temporibus et resistent."
40 *Ibid.*, f. 28r, c. 2. "Secundo premitto quod etsi inter complexiones magna diversitas sit, quelibet tamen est aliquante potentie in agendo et etiam in resistendo sive in conservando subiectum suum in dispositione sibi naturali."
41 *Ibid.*, f. 28r, c. 2.

gives a certain independence to the *complexio*. Now the fourth maintains that if the natural complexion of a body is intact, that body being in its natural disposition seeks to preserve this disposition when any attempt is made to alter it. This is true as long as the *complexio* is stronger than the factors seeking to alter it.[42]

In the example cited to make these suppositions clear, Marliani demonstrates his use of the word *temperamentum*, widely employed by the Arabic physicians from Galen. A body would seem to have *temperamentum* when it is in its natural condition of heat, health, etc. A body suffering from a fever would be spoken of as having *distemperamentum*. In this way the condition *temperamentatum* becomes a kind of goal or end of the complexions involved. When the complexions are too weak, the body becomes *distemperamentatum*.[43]

By the use of this conception of *complexio*, Marliani arrives at some additional conclusions. The first is a limited acceptance of John's theory of constant temperature. Many exterior, as well as some interior, members of healthy human bodies are equally hot intensively in the summer as they were in the preceding winter.[44] The complexion accounts for this. When in the summer environmental temperature is not too high, the *complexio* would remain intact, and tend to reduce the member to its natural temperature *(dispositio)*. The complexion works in just a reverse manner in the winter. At this time it has a greater quantity of *spiritus* to help it remain intact against the external cold.

42 *Ibid.*, f. 28v, c. 2, " Quarto premitto quod unumquodque corpus in naturalissima sua dispositione existens appetit in illa permanere et conservare atque si fuerit ab illa remotum, salvata sua complexione naturali, ad illam se reducet si amoveantur impedimenta hoc fieri prohibentia aut ita tantum minorentur ut complexio illorum naturalis conservata sit illis potentior.

43 See especially *ibid.*, f. 28v, for the examples illustrating the operation of the complexion.

44 *Ibid.*, f. 29r, c. 1, " Multa membra exteriora ac etiam aliqua interiora corporum humanorum robustorum... eque calida sunt in estate intensive sicut fuerint in hieme proxime precedenti... "

In certain circumstances, the second conclusion tells us, some members are colder intensively in the winter than in the summer, while some are intensively hotter. In support of the first part of this, he mentions the fact that fingers almost freeze in very cold weather.[45] For confirmation of the second part the theory of antiperistasis is cited. Because of the suddenness of the contact with the cold a great quantity of *spiritus* is recalled to the interior members. A union of *spiritus* occurs around the interior members, and the result is antiperistasis, or the intensification of the interior heat of some members above their natural disposition. Therefore, some members are hotter intensively in the winter than in the summer.[46]

Although he waits until the fifth part to discuss the methods of explaining antiperistasis, Marliani cites at this point several passages from the doubtful *Problemata* of Aristotle, the treatise of the same name attributed to Alexander of Aphrodisias, and the *Canon* of Avicenna to prove that antiperistasis takes place.[47]

The third conclusion mentions another category of members of healthy human bodies, both exterior and interior, which are colder intensively in the summer than in the winter preceding.[48]

The fourth conclusion [49] is an elaboration of the first, while the fifth tells us that it is not only naturally possible but often happens that some members of a given healthy human being become intensively heated in the course of the winter, and then afterwards during the same winter become intensively cooled. The example that he cites is of someone stepping out into the cold air. The *spiritus* recede to the interior members, and heat them above their natural heat. After a time, however, the *spiritus* will return toward the exterior members, reducing the interior members to their natural dispositions.[50]

45 *Ibid.*, f. 30r, c. 1.
46 *Ibid.*, f. 30r, c. 2.
47 *Ibid.*, f. 30r-v.
48 *Ibid.*, f. 31v.
49 *Ibid.*, f. 32r, cs. 1-2.
50 *Ibid.*, f. 32v, c. 2, "... satis est naturaliter possibile, immo fortassis sepe eveniens, quod aliqua membra dati hominis robusti intensive in aliqua parte

The sixth conclusion is of no importance, and the seventh and last is an extension of the first part of the second. It explains that a number of the members, mostly the extreme members, are found to be colder intensively in the winter.[51] The exterior porosity is more obstructed in the winter than in the summer, and this obstruction results from the actual cold state of the exterior members.[52]

In estimating Marliani's achievements, as represented by these premises and conclusions, we have on the credit side, a partial recognition with John Sermoneta of the constancy of body temperature. Furthermore in his favor was his realization that, after all, body heat involved the temperatures of individual members or organs. He roughly classed them as exterior and interior members, and questioned how they were affected by both environmental temperature and natural heat. To his credit, likewise, was his acceptance of the distinction between " intensively hot " and " naturally hot " already advocated by the earlier Italian physicians.

On the debit side we can list his identification of heat production with the generation of *spiritus* as too traditional. Nor was his apparent acceptance of antiperistasis in his favor.

Finally, however, in view of the difficulty today of understanding completely the mechanism of heat regulation, we must not be too scornful of his use of " complexions " for that purpose.

II. ANTIPERISTASIS

We have constantly come in contact in the first section of this chapter with a phenomenon called by the Greeks " antiperistasis ". We have seen that it was defined as the intensification of a quality by sudden contact with its contrary quality.

hiemis calefient que post in illa parte eiusdem hiemis intensive infrigidabuntur ... "

51 *Ibid.*, f. 33v, c. 2, " ... et plura membra et maxime exteriora intensive frigidiora reperiantur in hieme quam fuerint in estate immediate precedenti ... "

52 *Ibid.*, f. 34r, c. 1.

Diodoros credits Oenopides of Chios with originally observing that the waters of the Earth are cold in the summer and warm in the winter, that water in deep wells in the winter is not particularly cold, etc.[53] These observations, whether first made by Oenopides or not, are apparently the basis of the theory of antiperistasis. The word itself was first used by Aristotle, and through his authority remained current until at least 1665 when Robert Boyle made a new examination and criticism of the theory in his *New Experiments and Observations touching Cold, or An Experimental history of cold, begun. To which are added an Examen of Antiperistasis,* etc.[54] In this chapter we shall concern ourselves with some brief observations on the use of the theory by Aristotle and the attempted explanations of the theory made in the late Middle Ages.

Aristotle employs the theory chiefly in his *Meteorologica* in order to explain the production of hail and of rain in the summer, etc. The main difficulty in any explanation of the production of hail, according to Aristotle, is that although hail is ice, and water freezes in the winter, yet hailstones come primarily in the spring and autumn, sometimes in the late summer, but rarely in the winter.[55] The answer lies in antiperistasis. " Antiperistasis takes place between warm and cold. It is for this reason that in warm weather the lower parts of the earth are cold, while in very cold weather they are warm. The same thing, we must suppose, takes place in the region above (i. e. the air), so that in the warmer seasons the cold is concentrated (lit. " antiperistasized within ", ἀντιπεριστάμενον ἔισω) by the heat, causing the cloud to go over into water suddenly. . . .

53 K. Meyer, *Die Entwickelung des Temperaturbegriffs im Laufe der Zeiten,* Braunschweig, 1913, p. 3.

54 The treatment of antiperistasis by Aristotle, and in the modern period, with a special examination of this dialogue of Robert Boyle, is made by Kirstine Meyer in her " Zur Geschichte der Antiperistasis " in *Annalen der Naturphilosophie,* 3rd Vol., 1904, pp. 413-441. She, however, does not treat the period of the Middle Ages.

55 *Meteorologica,* I, 12, 348a.

But when the cold has been concentrated within still more by the outer heat, it freezes the water it has formed, and there is hail. We get hail when the process of freezing is quicker than the descent of the water. . . . Hail is rarer in summer than in spring and autumn, though commoner than in winter, because the air is drier in summer, whereas in spring it is still moist, and in autumn it is beginning to grow moist. . . . The fact that the water has previously been warmed contributes to its freezing quickly; for so it cools sooner. Hence many people, when they want to cool hot water quickly, begin by putting it in the sun. . . . It is for the same reason that rain falls in summer and not in winter in Arabia and Ethiopia, too, and that in torrents and repeatedly on the same day. For the antiperistasis due to the extreme heat of the country cools the clouds quickly." [56]

This rather lengthy passage from Aristotle has been included because it is the basis of practically all the arguments used by the medieval authors in support of antiperistasis, both those drawn directly from the *Meteorology* as well as those cited from the *Canon* of Avicenna, and the two treatises bearing the title *Problemata,* already mentioned in the first section.

The late medieval writers were not satisfied with merely citing the theory. They were interested in explaining it. We have seen that Jacobus de Forlivio approached the problem of body heat from two directions, the first of which was the physical one of antiperistasis.

The first thing to be noted in discussing antiperistasis, according to Jacobus, is that there are two ways which we can imagine one contrary invigorating or fortifying another. The first is by unifying or bringing together its parts so that these parts are more powerful for acting and resisting than they were before. The second is by intension in form.[57] The second would mean that in the case of a warm body being fortified by

---

56 *Ibid.,* I, 12, 348b-349a. I have altered slightly the E. W. Webster translation, Oxford, 1923.

57 Jacobus, *op. cit.,* ed. 1495, f. 10v, c. 2.

a cold the former is actually raised in temperature. In his next premise, Jacobus tells us that wherever heat or any quality is increased intensively or produced in a body, there must be some agent bringing about this intension or production, for it is inconceivable that the production of a quality could take place from a mere union of parts. As a further consideration Jacobus observes that one contrary never immediately and *per se* concurs in the production of another contrary.

Yet, in spite of all these premises and considerations, it is known that it is possible for one contrary surrounded by another to be intended (increased intensively), or in other words that antiperistasis takes place. Jacobus cites as examples the production of hail, and the more rapid freezing of heated than non-heated water.[58] How then can we solve antiperistasis? It cannot have taken place as a result of a first quality (i. e. the heat just can't produce itself). Nor can it have taken place directly from the contrary (i. e. the surrounding cold doesn't directly produce a higher heat by its own action). There must then be some other quality which is producing this antiperistasis or intension (increase in heat). This quality is simply the species or the reflected ray (radius) of the first quality (i. e. the heat) which is being intended (i. e. becoming hotter) in the subject. It is by reflexion, then, that antiperistasis takes place. However, it is not necessary that every quality surrounded by its contrary be increased intensively, for it is possible that the surrounding quality is more powerful in corrupting its contrary than are the reflected rays *(reflexi)* of the contrary in intending it. Naturally there would be no antiperistasis in such a case.[59]

This method of explaining antiperistasis was apparently the common one. Gaetan had used it in order to explain reaction. And, as Marliani now tells us, he had already criticized it in his first and second treatises on reaction.[60] Gaetan, we should re-

[58] *Ibid.*, f. 10v, c. 2, "... possibile est unum contrarium ab alio contrario circumdatum fieri in forma qua illi contrariatur intensius quam prius erat ..."

[59] *Ibid.*, f. 11r, c. 1, especially conclusions 1-4.

[60] Marliani, *De caliditate corporum humanorum*, f. 39r, 9v, c. 1.

member, had connected greater contrariety with greater reflexion,[61] and had declared in support of his thesis that there
was greater congelation in the middle region of the air in the
summer than in the winter because of antiperistasis and
reflexion. Marliani's early criticism of the theory rested upon
its inability to show how the middle region of the air can be
colder by reflexion. The cold in the middle region is of a certain
intensity, and the reflected cold, which derives from the original
cold of the middle region, cannot conceivably be colder than its
source. Therefore, it remains a mystery how a union of the
cold of the middle region with the reflected cold, which is of an
equal intensity, can produce the more intense cold needed to
explain antiperistasis.[62]

There were some who actually denied that any intensive increase took place in the phenomenon of antiperistasis. Marliani
in an earlier part of his *De caliditate,* while discussing the
views of a certain unknown doctor on body heat, mentions
such a theory. It maintained that although a body in which
antiperistasis was supposed to have taken place seems to be
hotter, yet actually it is not intensively hotter (i. e. of a higher
temperature). It is only that by the union of hot parts and the
better application, the hot parts help each other to act.[63] But
Marliani's answer to this method of explaining antiperistasis
is that experimentally you cannot find any part where this
union or concentration takes place. Water in a well in either
summer or winter seems to be uniform without any particular
concentration of heat.[64]

Marliani's position in this controversy over antiperistasis is
indefinite. He notes that he does not believe all the experiments
that are commonly narrated in support of antiperistasis, for

61 See Chapter II, *supra,* note 57 and discussion.

62 See Marliani, *De reactione,* Ms. cit., f. 39r-v, *In defensionem,* Ms. cit.,
ff. 4v-5v.

63 *De caliditate,* f. 9v, c. 1, "... etsi ad sensum percipiatur calidior, non
tamen est intensive calidior, sed per congregationem partium calidarum et
ad invicem meliorem applicationem ex quo partes sese iuvant in agendo."

64 *Ibid.,* f. 9v, c. 2.

instance that the air is hotter in subterranean places in the winter than in the summer. When he was a boy of thirteen, he firmly believed against the opinion of the more learned *(provectiorum)* that water only appeared to be hotter in the winter time to the exterior members which were colder themselves, rather than actually being hotter; and likewise in the summer that the water only appeared to be colder since the exterior members were warmer. This is rather shrewd judgment for a boy of thirteen. Now, however, at the time that he is writing, he does not think that this is the whole truth. But, he adds, he cannot deny that it might be, for he is not entirely sure *(omnino certissimum)* what is the truth in this matter.[65] At any rate, Marliani recites three methods of solving antiperistasis, confessing that he leans more *(magis adhereo)* to the first because its reasons are more efficacious *(efficatiores)* than those of the others.[66]

A quick summary of the first method will, I believe, cause us to wonder why Marliani did not conserve his youthful ideas rather than even partially support such a theory.

Like the *complexio* theory of body heat, this method suggests that a body has a tendency to conserve or to return to its natural disposition. The reduction of hot water to coldness after the heating instrument has been removed is presented in support.[67] Marliani is perfectly aware that this runs counter to the arguments he had presented against Johannes de Arculis, many years before, to prove that heated water is not cooled intrinsically.[68] Nor does he apparently change his view now to support this premise.

The method proceeds by a comparison with a theory of natural place, which remarks that although bodies in general tend to move to their natural place when that which is constraining them has been removed, it is possible that a body ac-

65 *Ibid.*, f. 49v, c. 1.
66 *Ibid.*, f. 49v, c. 2.
67 *Ibid.*, f. 39v, c. 2.
68 *Ibid.*, f. 48r-v.

cording to some particular part might be moved *ex se* farther from its natural place.[69] The example cited to prove this is that of the concave surface of water in a vase. Some parts of the surface are further removed from the center of gravity than they should be according to the strict interpretation of natural place.

The crucial assumption that similarly allows antiperistasis, or the intensifying of a quality beyond its natural disposition, is that a body in the presence of its contrary tends toward greater sphericity, or toward some shape in which its parts become more united.[70] He cites, in support of this assumption, an experiment of Buridan which tended to show that in the presence of cold air, the smoke (actually steam) in warm baths is more united.

When the body surrounded by its contrary has assumed its more efficacious shape as a result of that contrary, it becomes more powerful in acting and resisting and actually is able in some instances to increase itself intensively with respect to that quality according to which it is spoken of as a contrary, and thus antiperistasis is possible.[71]

It is difficult to see why Marliani prefers this method of explaining antiperistasis, especially when we realize that his

69 *Ibid.*, f. 40r, c. 1.

70 *Ibid.*, f. 40r, c. 2, "Premitto quarto quod propter unum corpus contrarium alicuius alteri esse approximatum contingit unum illorum fieri spericum aut ad spericitatem magis quam ante declinare sive ad figuram in qua eius partes magis quam ante fiunt unite..."

71 *Ibid.*, ff. 40v-41v. See the four conclusions. The fourth and its corollary declares: "Etsi non omne corpus a dato contrario circumdatum et exinde factum potentius ad agendum et resistendum quam pro certo tempore ante esset et quam sit contrarium circumdans se ipsum ultra intendet secundum qualitatem secundum quam est contrarium continenti, possibile est quod aliquod corpus ab alio dato contrario locatum se ipsum intendet secundum qualitatem secundum quam contrarium dicitur contrarium . . . (f. 42r-v) Correlarium, antiperistasis est hoc modo possibilis. Nam aliquod corpus iuxta aliquod contrarium positum se intendit secundum qualitatem secundum quam est ei contrarium..." The explanation that Marliani gives here for the fact that antipersistasis does not take place every time a body is surrounded by its contrary is that the natural disposition is often powerful enough to hinder or prevent the intensive increase that should follow.

previous denial of the intrinsic reduction of heated water brings into jeopardy the premise which allows a natural disposition to bodies. However, he seems to be impressed most with the experiments which tend to prove that a body is united in the presence of its contrary.

He remarks in one interesting passage that a vaporous body not only can unite in the presence of its contrary, but also can be moved locally. In support of this statement he alludes to a public experiment performed at Viglevani (Vigevano?) at the command of Galeazzo Maria, fifth duke of Milan. He speaks of many being present including his honored colleagues, Ambrosius Griffus and Tealdus Placentinus. Although Marliani does not tell us exactly what the experiment was, he remarks that the only way that he and his friends can explain it is that, " the fire being contrary to the cold water and so obviously fleeing from the cold water would be moved upward through the furnace." [72]

Perhaps an equally important reason for Marliani's preference for the first method was his feeling that the other common method of explaining antiperistasis, of which his so-called second and third methods are both variations, was even more untenable. This latter common explanation was substantially the reflexion theory as outlined by Jacobus de Forlivio and Gaetan of Tiene. In contrast to the first method which called for antiperistasis as a result of a more sufficient union of heat parts whereby a simple alteration of intensity takes place, this reflexion theory had as its *fundamentum* the belief that a first quality is generative of a *qualitas spiritualis,* which although different from a first quality was thought to be

---

72 *Ibid.,* f. 50r, c. 1, " Et etiam quia videmus manifeste corpus vaporosum non solum uniri in presentia contrarii ut magis se conservet sed etiam localiter moveri ab illo quemadmodum clare patuit in experimento facto viglevani iussu... Galeae Marie ducis quinti mediolani... coram multis presertim prestantissimis medicis et physicis Ambrossio Griffo et Tealdo placentino fratribus meis honorandis qui et ego non aliter experimenti illius effectum mirati salvare potuimus nisi in concludendo quod summus ignis contrarius aque frigide ab aqua frigida manifeste fugiens per caminum sursum moveretur."

representative of the first quality from which it was generated in so far as it would effect the organ of touch similarly. This *qualitas spiritualis* is spoken of as a " species " of the first quality with the property of suffering immediate multiplication through its medium.[73] It is in this manner that heat multiplies its species through a medium. However, in some bodies species of heat are multiplied (propagated) totally, or reflected totally, in others, partly multiplied and partly reflected.[74] When antiperistasis takes place, the intensification is the result of the reflection of species by the surrounding contrary.[75]

In discussing the final doubts raised against these methods of explaining antiperistasis, Marliani advances the possibility of ascribing the sudden production of hail, especially hail of enormous size, to the power of demons. This is a rare instance of Marliani's discussing the supernatural.[76]

When we attempt to estimate Marliani's study of antiperistasis, we are struck by his hesitance, the lack of the vigour and independence that marked his earlier works. He seems in this part of the work to be trying constantly to reconcile authority, while admittedly looking back to the time of his youth when he held a genuine disbelief in the possibility of antiperistasis.

73 *Ibid.*, f. 44r, c. 1, see second premise.

74 *Ibid.*, f. 44v, c. 1.

75 *Ibid.*, f. 45r, c. 2. The third method, which is practically the same, stresses the explanation of the more rapid reduction of heated water. It declares that the virtual frigidity of the water is reflected within by the presence of the contrary, and thus its power is increased, and a swifter reduction takes place. (f. 48r, c. 2) Marliani rejects this explanation, and provides an extremely detailed solution of the problem of more rapid reduction by premises and conclusions that attempt to prove that heated water and the non-heated water when exposed to the air are not of the same density, and as a result a difference in the rate of their freezing is to be expected. (ff. 51-55). In his early treatise when he confessed his ignorance of how such a more rapid reduction of heated water could take place, he had thought that there was no such difference in density. See Chapter III, *infra*, note 32 and discussion.

76 *Ibid.*, f. 55v.

CHAPTER V

THE UNIFORM EQUIVALENCE OF UNIFORMLY ACCELERATED MOTION

THE study in any mathematical detail of uniformly accelerated motion was something not accomplished by Antiquity, and for many years it was believed that such a study had been delayed until the advent of Galileo. But Duhem showed a generation ago that investigations into accelerated motion were made in the Middle Ages from at least the fourteenth century.[1] Two schools in particular are responsible for the early proofs of the fundamental theorem: that in equal periods of time one body moving with a uniformly accelerated velocity and another with a constant velocity equal to the mean between the initial and final velocities of the accelerated movement will traverse equal spaces. These schools are Oxford and Paris.

This theorem is especially important; for once it was understood the free fall of bodies was a uniformly accelerated movement, then the law of the free fall would follow immediately from the theorem in question. According to Duhem, it was the Dominican Soto, and not Galileo, who first applied this theorem to the free fall.[2] We are not concerned in this chapter, however, with the discovery of the law of the free fall, but only with proofs of the fundamental kinematic theorem.

The kinematic study of movement appears to have begun some time before 1328 with an anonymous treatise *De proportione motuum et magnitudinum,* variously attributed to Ricardus de Versellys (Vergellis, Uselis) and Gerard of Brussels.[3]

1 P. Duhem, *Études sur Léonard de Vinci,* Vol. III, Paris, 1913. See especially Part XV, Chapters XVIII, XXIII, *et passim.*

2 *Ibid.,* III, p. 560.

3 *Ibid.,* III, p. 295. G. Eneström, " Sur l'auteur d'un traite *De motu* auquel Bradwardine a fait allusion en 1328," *Archivio di storia della scienza,* Vol. II (1921), pp. 131-136.

Contrary to the impression that Duhem leaves, the further development of kinematics seems to have taken place in England before it did in France. Moreover, the law concerning the equivalence of accelerated and uniform movements was not discovered by Nicolas Oresme, as Duhem declares in one place, since it appears in the works of John of Dumbleton, Richard Swineshead, and Walter of Heytesbury, the first two of whom are without a doubt, and the last possibly, earlier than Oresme.[4] It is more than likely that the theorem in question, discussed as it was in England before 1350, passed from Oxford to Paris. That the schools were in close contact would seem to follow from a Parisian student's summary of the English ideas dated about the middle of the century, or perhaps later (Paris, BN Ms. Fonds latin no. 16621).

It will develop as the chapter proceeds, and the various proofs of the English and French schools are alluded to, that with Oresme the French school tends by the use of coordinates toward a more geometrical proof, the English of Heytesbury, Dumbleton, and Swineshead toward one more arithmetical and formally logical. And while the Italians of the fifteenth century, the offsprings of both of the earlier schools, seem to lean more heavily on the English logicians, still they were certainly not unaware of the coordinate system of Oresme (see note 25 *infra*). Marliani follows the general course of Italian development, his main inspiration being English, direct references being made to the Calculator and Dumbleton, both of whom we shall have occasion to discuss later.

As we have noted in the first chapter (See Part II, Section E), Marliani's short treatise which deals with accelerated motion is entitled *Probatio cuiusdam sententiae Calculatoris de motu locali*. Written in 1460, it is Marliani's first effort on the subject of motion. The study opens with a brief explanation as to why it has been written. Marliani tells us that in proving

4 Duhem, *op. cit.*, III, p. 389. Dumbleton can be dated at Oxford between 1331-1349. For a discussion of Swineshead as the author of the *Liber calculationum*, see the appendix. For Heytesbury, consult Chapter II *supra*, note 16.

UNIFORMLY ACCELERATED MOTION

that latitude of motion uniformly difform corresponds to its mean degree, the Calculator in the third proof stated a consequence that might be denied or doubted by many.[5] As a result then of possible ambiguity in the Calculator's proof, Marliani decides to present his own demonstration of the theorem, that with respect to the space traversed in a given time, a latitude of motion uniformly difform (i. e. a uniform change in speed or velocity) can be represented by its mean degree (its mean speed or velocity).

Concerning a latitude of local motion uniformly difform (uniform change in velocity), or in fact concerning a latitude (alteration) of qualities uniformly difform in general, there were, according to Marliani, two prevailing opinions. The first of these held that any latitude of motion uniformly difform " in defining the subject " corresponds to its most intense degree.[6] Now it was quite clear to Marliani and all those who accepted the theorem of the mean velocity that one way the subject (the moving body) can be " defined " in the case of local motion uniformly accelerated is by the amount of space that a moving body can traverse in a given time. So that when an equivalent uniform movement is sought to represent uniformly accelerated movement, the common basis of definition is the quantity of space described in the same period of time. Whether this " definition " or " denomination " of the subject was used by the supporters of the " theory of the most intense degree " [7] (i. e. those who believed the accelerated velocity

5 Venice, San Marco, VI, 105, f. 8r, c. 1, " Quoniam Calculator in probando latitudinem motus uniformiter difformem suo gradui medio correspondere in tertia probatione format unam consequentiam ... que fortassis multis esset dubia aut a multis negaretur, cum dicit sequitur quod c equaliter distat a 4or, sicut d a non gradu, et ergo 6 equaliter distat a c, sicut duo a d, et cetera ... " This is a specific point in the Calculator's proof, and is therefore unintelligible until the proof as a whole has been studied. For the Calculator's third proof containing this particular passage, see note 19 infra.

6 Ibid., f. 8r, c. 1, " ... quelibet talis latitudo in denominando subiectum suo gradui intensissimo correspondeat."

7 This opinion is adopted by Roger (Richard? See Appendix I) Swineshead in his De motibus naturalibus et annexis (Erfurt, Stadtbücherei,

was defined by its final velocity) we can not always determine from their often ambiguous statements.

The second of the prevailing opinions that Marliani mentions is that of the mean degree, namely that a latitude of motion uniformly difform (uniform change in velocity) corresponds with respect to the amount of space described in a given

Amplonian F 135, f. 38r-v; Paris, BN Fonds latin, no. 16621, f. 62r: *cf.* Duhem, *Études*, III, p. 454). It is the subject of refutation in the *Tractatus de sex inconvenientibus* (edition included with Bassianus Politus, *Questio de modalibus*, Venetiis, per Bonetum Locatellum Bergomensem, 1505). (Also ms. in BN Fonds latin, no. 6559). He tells us: " Propter illa et similia argumenta... dicitur a quibusdam quod in latitudine motus localis terminata ad non gradum tota latitudo motus non est equalis suo gradui medio; nec sibi correspondet, sed solum gradui intensissimo, sic quod denominatio latitudinis totius sit a denominatione gradus intensissimi in illa latitudine... sed hoc totum est falsum ". (*ed. cit.*, f. 55r, c. 1; *ms. cit.* f. 37v, c. 2). Duhem (*op. cit.*, III, p. 487) assigns this viewpoint also to Jacobus de Forlivio on the authority of Louis Coronel (*Physicae perscrutationes*, Paris, 1511, f. LXVI, c. 1). Duhem asserts that he trusts the accuracy of Coronel's statements, though he has not been able to read Jacobus himself. Coronel declares that Jacobus holds that a latitude is equivalent to the most intense degree within it. (Coronel, *op. cit.*, Paris, 1530, f. 73v, c. 2). The position of Jacobus is rather difficult to determine. In one place I have noticed that in adducing some arguments against a conclusion of Walter Burley in his *De intensione et remissione* (Venice, 1496, f. 20r, c. 1) Jacobus says: " cum igitur *d* latitudo sit uniformiter difformis requiritur quod ipsa correspondet suo gradui medio ..." He clearly upholds the common view here. In another passage Jacobus seems to hold the opposite theory, but his presentation there of the correct theory is far better than his answers to that theory. The form is so highly scholastic that it is sometimes difficult to determine just what Jacobus himself thinks. The passage (Venice, 1496; f. 36r, c. 1 ff. Undated ed. Pavia? f. 52r, cs. 1-2) is as follows:

In a previous argument he had used as fundamental " quod quilibet motus difformis secundum se totum ad aliquem gradum terminatum secundum extremum intensius est ita intensus sicut ille gradus ad quem terminatur." But he admits that there is an opinion that would deny this and substitute instead " quelibet motus uniformiter difformis quantum ad tempus est precise ita intensus sicut gradus medius eiusdem. Igitur non est ita intensus sicut extremus gradus ad quem terminatur. Consequentia patet et antecedens arguitur; quia per quemlibet tantum motum in equali tempore precise equale spacium describitur sicut per motum uniformem sub eius gradu medio. Capiatur enim motus uniformiter difformis a non gradu ad gradum ut octo et vocetur *a* et eius medietas intensior sit *b* et remissior *c* et capiatur *d* motus uniformis sub gradu ut 4 qui est medius et cetera. Et patet quod

time to its mean velocity or degree (gradui media per eque distantiam eius extrema). We are told that this is the opinion of the Calculator in his *De difformibus* [8] and also that held by

quantum spacium pertransibitur per *a* tantum pertransibitur per *d* per equale tempus. Quod arguitur sic: quia precise per quantum *b* medietas intensior ipsius *a* excedit secundam medietatem *d* tantum prima medietas *d* excedit remissiorem medietatem *a* scilicet *c* igitur totum est equale toti. (This is similar to the reasoning of Oresme, but is not translated geometrically; see note 25 for Oresme's proof)—consequentia patet per hoc: quia quandocumque aliquorum duorum resultantium ex duabus partibus partes habent se sic quod quanto una illarum partium excedit alteram partem alterius totius, tantum econverso alter pars exceditur ab altera parte; ista sunt equalia." After a further exposition of this position that holds a latitude can be represented by its mean degree, Jacobus proceeds to answer it in two ways. The first method of responding, the only one in which we are interested here, declares that latitudes are equivalent to their most intense degrees.

This theory of the "most intense degree" quite possibly was derived from the theory of rotary motion made popular by Bradwardine, and accepted by the whole English school, "the velocity of local motion follows the velocity of the most rapidly moving point in the body locally moved" (See Bradwardine, *Proportiones*, Bibl. Nat. Ms. Fonds latin, no. 6559, f. 36v: "...velocitas motus localis attenditur penes velocitatem puncti velocissime moti in corpore moto localiter ..."). At least Bradwardine is often cited in support of the theory of the final velocity. But Bradwardine is speaking here of the distribution of velocity in space, not time. Although this was not always well stated, we may suppose that a similar difformity of distribution in space was understood by the supporters of the theory of the most intense degree, and that their position was misinterpreted by the followers of the theory of the mean degree.

8 In his *Liber calculationum* (Padua, Johannes de Cipro, 1477?) Swineshead treats the problem in several ways both in the section *de difformibus*, which Duhem in his *Études* (III, p.. 479-80) has described and in the section *de motu locali*. The third of those in the latter section is the one with which Marliani is chiefly concerned; and we reserve that for comparison under note 19. The first method in the *de motu locali* involves first proving that the composite of two unequal degrees is double their mean. Then he poses two bodies *a* and *b*, one acquiring latitude at the same rate as the other is losing it, so that the same amount of latitude is acquired as lost. He lets *c* be the mean of the latitude, and shows that *d*, the composite of *a* and *b*, is continually equal to twice the mean by his first supposition, and since over the whole period of time *a* will be equivalent to *b* (i. e. acquiring as much latitude as the latter lost), either one of the motions in the whole period of time is equivalent to the mean *c*. The pertinent passage is as follows: (*ed. cit.*, 55v, c. 1), "omne compositum ex duobus inequalibus est duplum ad medium inter illa ponatur quod *a* sit maius *b*, et sit *c* medium inter illa et diminuatur

Marliani in annotating both the Calculator's work and a *De instanti* of Peter of Mantua.[9]

It is this latter opinion that Marliani sets out to prove. He lays down first the fundamental suppositions with their proofs and then proceeds directly from them to the important conclusions. His first supposition is one, he tells us, on which all the treatises on the subject agree, namely that the latitude of local motion uniformly difform or in fact difform in any way with respect to the time employed for the traversal of space, corresponds to some intrinsic degree in that latitude, which is obviously more remiss than the more intense extreme and more intense than the more remiss extreme ($V_F > V_m > V_0$ where V_F is the final velocity [most intense degree], V_0 is the in-

---

a ad *c* et maioraretur *b* equevelociter versus *c*, tunc in fine compositum ex *a* et *b* erit duplum ad *c*, eo quod *a. b.* inter se tunc erunt equali. Sed compositum ex *a. b.* continue erit tantum sicut in fine (notice here rudiments of the method of the infinitesimal calculus) quia quantum unum acquiret tantum aliud deperdet. Ergo compositum ex *a. b.* nunc est duplum ad *c* quod fuit probandum... primo acquirat *a* aliquam latitudinem uniformiter terminatam in extremo intensiori ad *c* gradum, et deperdat *b* eandem latitudinem consimiliter omnino sicut *a* acquirit illam. Et sit *c* gradus medius illius latitudinis, ... tunc *ab* motus continue equivalebunt gradui duplo ad *c* eo quod gradus compositus ex duobus inequalibus semper est duplus ad medium inter illos et *c* semper erit medius inter *a. b.* quoniam erunt inequales et illa duo coniuncta continue eidem gradui correspondebunt eo quod unus equevelociter intendetur sicut alius remittetur. Ergo continue correspondebunt gradui dupli ad *c* et unus eidem gradui correspondebit sicut alius. Ergo illa duo gradui in duplo intensiori correspondebunt quam unus eorum per se. Ergo motus utriusque per se *c* gradui in totali tempore correspondebit quod fuit probandum. Consequentia tenet. Nam moveantur *d* gradu duplo ad *c* et sequitur quod *ab* pertransibit tantum *d*, et unum tantum pertransibit sicut aliud. Ergo unus pertransibit in duplo minus quam *d* et per consequens tantum sicut moveretur *c* gradu subduplo ad gradum quo movebitur *d.* qui gradus est medium latitudinis acquirende..."

9 In referring to annotations on the Calculator's *De difformibus*, Marliani is probably speaking of his *Tractatus physici*, the section entitled "De intensione et remissione in difformibus" (R. Bibl. Univ. di Pavia, Aldini codex 314, f. 4v, ff.). I have not found any work of Marliani's annotating the *De instanti* of Peter of Mantua (See Chapter I *supra*, part II, section D). However, the *De instanti* itself assumes that a latitude's mean defines it (Columbia University Ms. X 570 P 44, f. 4r) : "Et idem arguitur si ponantur diformes sive ponantur illos correspondere gradibus eorum mediis..."

itial velocity [most remiss degree], and V_m is the uniform velocity by which the change in velocity is to be represented).[10]

For the proof of this first supposition, he posits a body a moving with uniform acceleration (uniformly difform motion) for an hour through the distance b. There must exist some uniform velocity with which another body could move through the distance b in an hour, because there can be a given degree of uniform motion with which a body moves through a distance less than b, and (as is added in the margin of F. 8r c. 1) there can be a given degree of motion with which a body moves through a distance greater than b. Therefore since a degree of motion intended *ad infinitum,* or even remitted *ad infinitum,* can be received, there is a degree of motion c with which a body d moving with uniform velocity will traverse the distance b. (I have been using the expression velocity and speed interchangeably, since the motion considered here is rectilinear with no change in direction). It follows, Marliani continues, that this c degree of motion is not as intense as that degree which terminates the latitude with which a is moved in its more intense extreme, nor is it as remiss as that latitude's more remiss extreme (i. e. $V_F > V_m > V_0$). This conclusion is explained thus. If c degree of motion were just as intense as that degree with which the latitude of a is terminated at its more intense extreme, the following impossible situation would result from the 6th book of the *Physics* through the definition of " the quicker," [11] namely that for any given instant of the hour in which a is moved, as was posited in the case above, and also in which d is moved, d will move more rapidly than $a,$ and yet will traverse no more distance in that hour than a. On the other hand, there would be a similar impossible situa-

10 *Ms. cit.,* f. 8r, c. 1, "... latitudo motus localis uniformiter difformis, aut etiam quomodocumque difformis secundum tempus quantum ad pertransitionem spacii, alicui gradui intrinseco in illa latitudine qui scilicet est remissior extremo intensiore illius latitudinis, et intensior remisisore correspondet..."

11 *Physica,* VI, ii, 232a, 25-27, "... necesse est velocius in aequali tempore maius, et in minori aequale, et in minori plus moveri; sicut definiunt quidam ipsum velocius."

tion if *c* were that degree which terminated the latitude of *a* at its more remiss extreme, for then *a* would be constantly more intense than *d*, and yet would traverse no more space in the hour. So the conclusion is that *c* degree is more intense than the more remiss extreme of the latitude of *a,* and more remiss than the more intense extreme of that latitude, which of course is nothing other than saying that the degree *(c),* with which a body moving uniformly describes the same distance in the same times as does a body which acquires a given latitude in that time, is intrinsic in that latitude.[12] At this point it is sufficient to have proved that the degree *c* is intrinsic; that, in addition, it is equal to the mean degree of the latitude of *a* is reserved for later proof.

Marliani now passes on to his second general supposition, that if two bodies begin to acquire a form from no degree, i. e. a form of which there is a privation in those bodies, and if they acquire it uniformly and continually in a given proportion, one more swiftly than the other; for any chosen time, the one which acquires the form more swiftly than the other will have acquired more of that form (a greater latitude) according as the fixed proportion by which it acquired the form more rapidly. This follows from our definition of acquiring latitude more rapidly (or if we were speaking of motion, we would say by our definition of acquiring latitude of motion more rapidly, i. e. our definition of acceleration).

This general supposition is then applied to local motion. Two bodies are posited, *a* and *b*: *a* moves twice as rapidly as *b,* and does so constantly. Therefore, at any chosen instant *a* will have acquired twice the latitude of motion that *b* has. (Marliani means that if the acceleration of a body *a* is twice that of a body *b,* then at any moment t_1, its change of velocity

12 *Ms.* cit., f. 8r, 2. "Non est aliud dicere, quod ille gradus motus, quo uniformiter movendo per tantum tempus, in quanto mobile acquiret datam latitudinem motus uniformiter, tantum pertransiretur spacium quantum a mobili sic acquirente uniformiter illam latitudinem motus, est intrinsicus in illa latitudine."

from some time t_0 will be twice the change of velocity of b in the same period).

If we take two bodies moving as in supposition two, then a third supposition is evident, namely that if two bodies are moving in such a way that one is acquiring latitude of motion uniformly with more swiftness than the other, but in a fixed proportion for every point in the given time, the body moving more rapidly will traverse more space than the slower body in the given time, and in the same fixed proportion as their relative accelerations.[13]

From this supposition, it is evident that the degree of motion (velocity) to which the body which is moved more swiftly corresponds, i. e. in moving an equal space in an equal time, is in the same fixed proportion to the degree of motion to which the latitude of the more slowly moving body corresponds, as are the latitudes, both latitudes having the initial degree of zero.[14]

This follows from the fact that the degree of motion to which the latitude corresponds was made by definition that which traverses an equal space in an equal time; for which fact Marliani refers us to the first supposition.

In the following, or fourth, supposition, the author declares that any latitude of motion uniformly difform corresponds to an intrinsic degree of that latitude with respect to the quantity of space traversed, and that it thus corresponds to that degree whether it has been acquired in a short or long period of time. His proof is one based on proportions. Before actually making

13 *Ibid.*, f. 8v, c. 1, " 3° Presuppono quod si 2° (secundo) mobilia pro dato tempore, ita movebuntur quod in data certa proportione pro quolibet instanti illius dati temporis, unum datum altero continuo velocius movebitur, in illo dato tempore in illa certa data proportione maius spacium pertransibit velocius motum, quam tardius motum, ceteris paribus."

14 Assuming that in a given time $\Delta V_1/\Delta V_2 = K$, then $V_{m_1}/V_{m_2} = \Delta V_1/\Delta V_2 = K$ where V_{m_1} and V_{m_2} represent the uniform velocities to which the changes in velocity, ΔV_1 and ΔV_2, corresponds, when the bodies suffering ΔV_1 and ΔV_2 accelerate from rest.

the proof, he restates the problem. If a latitude of motion
from 2 to 6 degrees is acquired by a uniformly in an hour (the
speed of a body is uniformly accelerated from a velocity of
2 to one of 6), and by b uniformly in two hours (or in any
longer time), then the contention is that it (the latitude) corresponds to the same degree with respect to a and to b. If the
total latitude is acquired by a in one half the time it is acquired
by b, then any part of it is acquired by a in one half the time
it is by b. It does not appear therefore that the latitude should
correspond to a more intense degree with respect to a than to b,
or to a more remiss degree with respect to a than to b. Marliani
then demonstrates his contention by the following case posited
(the algebraic expression in parentheses are my own and not
Marliani's) : g is a latitude acquired by a and b uniformly in
unequal times c and d respectively. Let $\dfrac{d}{c}$ be in f proportion.
Let e and h be the spaces described by a and b in times c and d.
Then the author shows that the spaces traversed by a and b
are in the same proportion as the times $\left(\text{or that } \dfrac{d}{c} = \dfrac{h}{e} = f \right).$
He resorts to infinitesimals to prove this : a and b will acquire
each succeeding degree of the latitude in two instants which
are in the same proportion as the total times c and d. Then if c
and d were equal, in the two given instants a and b would
traverse equal spaces. But since c and d are not equal, a and b
must traverse in the given instants distances in the same proportion as c and d. This would be true for the whole latitude
acquired by a and b, so that the distance e traversed by a in time
c is in the same proportion to that time c as the distance h tra-
versed by b in time d is to that time d $\left(\text{i. e. } \dfrac{e}{c} = \dfrac{h}{d} \right).$ By definition the degree to which a latitude corresponds is that which de-
scribes the same distance in the same time. Then $\dfrac{e}{c}$ is the
degree to which the latitude g corresponds when acquired by

a in time $c;$ and $\dfrac{h}{d}$ is the degree to which the latitude g corresponds when acquired by b in time d. But $\dfrac{e}{c} = \dfrac{h}{d}$. In other words, a latitude of motion uniformly difform is equivalent to the same intrinsic degree whether it is acquired in a long or short period of time.

An important part of the proof of this fourth supposition was the statement with regard to the relation of the time and distance under the conditions posited. Out of it comes the proposition that if two bodies acquire the same latitude in unequal times, the distances traversed by those bodies are in the same proportion as the times (i. e. $\dfrac{S_2}{S_1} = \dfrac{T_2}{T_1}$, when S_2, S_1 are the distances, T_2, T_1 the times).[15]

Using these four previous suppositions and their corollaries, Marliani proves a fifth and very important one. Every body which accelerates its motion uniformly from rest or from no degree of motion through any given time will traverse three times as much space in the second half of the given time as in the first.[16] This supposition has been posited and proved by many, says Marliani, especially by the Calculator, who did so briefly and clearly, and although they are not mentioned by Marliani, Oresme and Heytesbury also give us proofs of it.[17]

15 *Ms. cit.*, f. 8v, c. 2. " Ergo si duo mobilia a quiete aut equalibus gradibus motus in temporibus inequalibus acquirunt eamdem latitudinem motus, qualis est proportio temporis ad tempus, talis est spacii ad spacium..."

16 *Ibid.*, f. 8v, c. 2. " Omne mobile quod a quiete sive a non gradu motus, per aliquod datum tempus, uniformiter intendet motum suum, in 3° (triplo) maius pertranscibit spacium in 2ª (secunda) medietate dati temporis quam in prima..."

17 The Calculator brings this up as the most important part of his proof. The passage is as follows: (*ed. cit.*, f. 55v, c. 1-2) " Et primo arguitur quod quecumque potentia a non gradu uniformiter intendit motum suum, in triplo plus pertransibit in secunda medietate temporis quam in prima. Ponatur enim quod *a* remittat motum suum uniformiter ad non gradum. tunc in omni instanti prime partis proportionalis in duplo velotius movebitur quam in instanti correspondente secunde partis proportionalis et sic deinceps ut patet. Ergo cum prima pars proportionalis temporis sit dupla ad secundam patet quod in quadruplo plus pertransibit in prima parte proportionali quam in secunda, et in quadruplo plus in secunda quam in tertia et sic in infinitum.

But Marliani's proof proceeds from the above suppositions:
d acquires latitude from 0 to 8 degrees in two hours; a acquires
latitude from 0 to 4 in one hour; b from 0 to 8 in one hour;
and c from 0 to 8 in two hours (as does d). Finally, we let a
move a distance of a foot in one hour. As a result: (1) When
compared with a, b moves two feet in an hour. This follows
from suppositions two and three. $\left(\dfrac{\Delta V_1}{\Delta V_2} = \dfrac{S_1}{S_2},\right.$ substituting
values, $\left. S_2 = 2.\right)$ (2) When compared with b, it is clear from
the fourth supposition (see Note 15 *supra* and text) that c
moves a distance of four feet in two hours. $\left(\dfrac{S_1}{S_2} = \dfrac{T_1}{T_2}\right.$ substituting values, $\left. S_1 = 4.\right)$ (3) c in the first hour will move only

Ergo in omnibus partibus in quadruplo plus quam in residuis partibus a
prima consequentia tenet per hoc: si alique quantitates ad alias comparentur
in eadem proportione sicut unum illorum se habet ad sui partem ita con-
gregatum ex omnibus comparatis se habet ad omnia ad quem sit comparatio
sequitur ergo in toto tempore in quadruplo plus pertransibit quam in secunda
medietate. Et per consequens in prima medietate in triplo plus transibit quam
in secunda..." Oresme in *De figuratione potentiarum et mensuarum dif-
formitatum*, referring to a system of coordinates (see note 25), includes
a somewhat different statement for proportionally varying periods of
time, namely: "If some period of time has been divided into proportional
parts; so that in the first part it moves with a certain speed, in the second
part with twice that speed, in the third part with thrice that speed, and so
in the end, the speed continuing to increase always in the same manner, this
speed would be exactly quadruple the altitude of the first part,... in the entire
hour a moving body would traverse four times as much space exactly as that
which it has traversed in the first proportional part, this is to say in the first
half hour, if, for example, in this first proportional part it has traversed a
distance of one foot, during the remainder of the time, it will traverse three
feet; and during the whole time a distance of four feet." (Quoted through
Duhem, *op. cit.*, III, p. 393.) Heytesbury is one of the first to advance the
proposition, but he does so after he has assumed that a latitude of motion
uniformly difform is equivalent to its mean degree. In his *Tria predicamenta*
(in edition of his works, Venetiis, 1494, f. 40v) : "Cum enim a non gradu ad
aliquem uniformis fiat alicuius motus intensio subtriplum pertransibitur precise
in prima medietate temporis ad illud quod pertransibitur in secunda, et si alias
ab eodem gradu aut ab aliquo quocunque ad non gradum uniformis fiat re-
missio triplum precise pertransiretur in prima medietate temporis ad illud
quod pertransiretur in secunda..."

one foot, since in that first hour it will have the same latitude as a. (4) Since the conditions of d are those of c, d will traverse 4 feet in the two hours, and only one foot in the first hour, or hence, three times as much distance in the second as in the first. And thus the proposition has been demonstrated.

Finally, we have reached from these five suppositions the last or sixth supposition and its corollaries. Any body which is moved in a given time with a given degree of motion will traverse so much space in that time as two bodies, one of which is moving in that time with a part of that degree, and the other of which is moving in that time with the remaining part of that degree; if, of course, the spaces traversed by these two bodies are joined together.[18] Again Marliani resorts to proportions for his proof. Let a move with 8 degrees of motion (velocity) for one hour a distance of 4 feet, b with a velocity of 6 for the same time, and finally c with a velocity of 2, also for one hour. Since we know from the definition of velocity that when the time is constant: $\dfrac{V_1}{V_2} = \dfrac{S_1}{S_2}$, the proof of this supposition becomes:

(1) $\dfrac{Va}{Vc} = \dfrac{Sa}{Sc}$ and filling in values $S_c = 1$ foot.

(2) $\dfrac{Va}{Vb} = \dfrac{Sa}{Sb}$ and filling in values $S_b = 3$ feet.

(3) $\therefore S_c + S_b = 4$ feet $= S_a$; " quod fuit probandum ", says Marliani.

The first corollary of this sixth supposition is that the space described by a body which acquires latitude of motion uniformly from one degree to another (accelerates from one velocity to another) is equal to the spaces described by two bodies, one of which is moving uniformly with the degree

18 *Ms. cit.*, f. 9r, c. 2. " Presuppono quod quodlibet mobile quod per datum tempus dato gradu motus movebitur, pertransibit tantum spacium in illo dato tempore, quantum spacium in illo dato tempore pertransirent duo mobilia, quorum unum moveretur per illud datum tempus una parte dati gradus motus, quecumque pars sit illa, et aliud moveretur residua parte dati gradus motus per datum tempus, horum duorum mobilium spacia in directum coniungendo."

(velocity) equal to that of the more remiss extreme (initial velocity) of the latitude (change in velocity) of the original body, and the other of which is acquiring latitude uniformly from no degree (o) to that degree equal to the difference between the final and original degrees of the latitude of the first body (fol. 9v, c. 1). The case that Marliani cites to prove this corollary is of body a which is acquiring latitude uniformly (i. e. accelerating uniformly) from 4 to 8 degrees of motion (i. e. from a velocity of 4 to one of 8) in an hour. Then it is necessary for him to prove that the space traversed by a in the hour is equal to the sum of the spaces described by a body b moving uniformly with a velocity of 4 and a body c which acquires latitude of motion uniformly from o to 4 degrees (the difference in the terminal degrees of a), both bodies moving for an hour. It is quite obvious, according to Marliani, that the velocity of a at any instant whatsoever (pro quolibet instanti) of the hour is composed of that of b and the velocity of c (i. e. 4 plus the velocity of c). Therefore, from supposition six it follows that at that instant as much space would be traversed by a as by b and c together. This condition holds true for any instant whatsoever of the hour. Since this is true, it would follow for the whole hour (or in other words Marliani appears to assume the truth of a summation process), and a during the given time covers as much space as b and c. We understand in addition that the latitude of a corresponds with respect to the amount of distance traversed in a given period of time to the degree of motion of b plus the latitude of c. Since we know from the first supposition that a latitude (or change in velocity) can be represented (with respect to space traversed in some given period of time) by some uniform degree (constant velocity), the second corollary to supposition six follows, namely that a, moving as posited above, will traverse as much space as b, moving as above, and another body, moving with a constant velocity equal to the mean velocity of the latitude of c, as given above (compare this discussion with the recapitulation at the end of the chapter).

The author is now ready to prove that the latitude of motion is defined by its true mean. He does so first for a body moving with an initial velocity of zero, or a body originally at rest. Let *a* acquire latitude of motion uniformly from 0 to 4 degrees in the first hour (i. e. accelerate uniformly from a velocity of 0 to one of 4), and from 4 to 8 in the second hour. Let *b* move in the second hour with a uniform degree of motion such as 4. Finally, let *c* move in the second hour with a uniform unknown degree *d* which would represent the latitude acquired uniformly by a body moving in the second hour from 0 to 4 degrees. (1) Then from the second corollary of the sixth supposition, *a* will move a distance equal in the second hour to the combined distances of *b* moving with 4 degrees uniformly and *c* moving uniformly with *d* degree. (2) If *d* degree is not the true mean of the latitude acquired uniformly from 0 to 4 degrees; it is, therefore, either more intense or more remiss than that mean. If it be more intense, then the extreme 4 is not double *d*; for, of course, it is double the mean, 2. (3) Then the degree composed of 4 plus *d,* which is the degree by which the latitude of *a* acquired uniformly in the second hour from 4 to 8 is represented is not thrice *d,* since 4 is not twice *d.* (4) *d* which is the degree defining the latitude uniformly acquired from 0 to 4 by *c* in the second hour is also the degree to which the latitude acquired by *a* in the first hour corresponds. (5) But since 4 plus *d,* the equivalent of the latitude of *a* in the second hour, is not thrice *d,* then the distance traversed by *a* in the second hour is not thrice that traversed in the first hour. But this conclusion is opposed to supposition 5; and hence *d* is *not more intense* than the mean, 2. (6) To continue, if *d* is more remiss (i. e. less) than the mean, then 4 is more than double *d.* (7) Therefore 4 degrees contains 2*d* (i. e. twice *d*) plus some additional degree; and (8) the degree of motion composed of 4 plus *d* contains 3*d* plus an additional degree. (9) Hence, 4 plus *d* will traverse more than three times the distance of *d.* But this conclusion is also opposed to supposition 5. Consequently, *d* is *not more* *remiss* than the mean, 2. Since *d* is neither more intense than the mean, nor more remiss, and yet is intrinsic, therefore it must be equal to the mean.

This indirect proof should be compared with that of Swineshead.[19] It is in this proof of Swineshead that we find the original statement, the possible denial of which caused Marliani to compose his treatise. The chief difference between the proof of the Calculator and that of Marliani is that there is no supposition like the sixth and its corollaries in the Cal-

19 *Ed. cit.*, f. 55v, c. 2. "... Intendat enim *a* uniformiter a non gradu usque ad octo. tunc si motus in prima medietate correspondebit gradui remissiori quam duo, sit ille gradus *d*. et sit gradus cui correspondebit motus eius in secunda medietate *c*. Tunc ex quo motus equevelociter intendetur in secunda medietate sicut in prima et instanti medio habebit quatuor. Sequitur quod *c* equaliter distat a quatuor sicut *d* a non gradu et sex equaliter distat a *c* sicut duo a *d*, igitur sex equaliter distat a duobus et sicut *c* a *d*. Ergo cum sit tripla proportio sex ad duo et gradus *c* est remissior sex patet quod inter *cd* est maior quam tripla proportio. Et per consequens plus quam triplum pertransibit in secunda medietate quam in prima. Consequens est improbatum. Et quod sequitur *c* equaliter excedere *d* sicut sex duo patet per hoc quod si sint quatuor termini quorum primus et maximus equaliter excedit secundum sicut tertius quartum, primus equaliter excedit tertium sicut secundus quartum ... (then follows this statement in terms of *a, b, c* and *d*; i. e. if *a* minus *b* equals *c* minus *d*, then *a* minus *c* equals *b* minus *d*.) ... tunc distantia inter *a.c.* est maior distantia inter *a.b.* solum per distantiam inter *b.c.* et excessus *b* supra *d* est maior quam excessus *c* supra *d* solum per latitudinem inter *b.c.* Ergo cum inter *a.b.* et *c.d.* est distantia equalis, patet quod inter *a.c.* et *b.d.* est distantia equalis quod est probandum. Sic ergo motus in prima medietate gradui remissiori medio correspondebit, patet quod motus in secunda medietate plus quam in triplo intensiori gradui correspondebit quam in prima. Consequens est falsum et improbatum. Si motus gradui intensiori medio correspondebit sequitur quod cum motus in secunda medietate equaliter excedit motum in prima medietate sicut sex excedunt duo. Et patet quod motus in secunda medietate non foret triplus ad motum in prima. Consequens est falsum per prima argumenta. Patet ergo quod motus in prima medietate correspondit gradui ut duo et in secunda gradui ut sex et sic arguitur de omni latitudine ad non gradum terminata. Idem sequitur si intendat a gradu quia acquiret latitudinem de novo a non gradu que correspondet suo gradui medio et excedit gradum a quo intendet per latitudinem istam de novo acquisitam et gradus medius inter gradum in principio habitum et gradum in fine habendum est intensior illo gradu nunc habito per medium totius latitudinis de novo acquirende, ergo sicut illa de novo acquisita correspondet medio gradui, ita cum toto motu prehabito suo gradui medio correspondet quid fuit probandum."

culator's (but such a supposition is included by John Dumble-ton). They were put in Marliani's proof to clear up any doubt arising from the earlier proof. Likewise both proofs depend largely on what appears as Marliani's fifth supposition, concerning the comparative distances traversed in the first and second half of the time involved.

Having proved it indirectly, Marliani uses the same supposi-tions to prove directly that the latitude (change in velocity) of a body accelerating uniformly from rest corresponds with respect to the space traversed in the same time to its mean (velocity). This time a acquires latitude uniformly in the first hour from 0 to 8, and from 8 to 16 in the second hour; b moves uniformly with a degree of 8 in the second hour; while c in the second moves uniformly with a degree d equiva-lent to a latitude acquired from 0 to 8. The latitude of motion of a in the second hour is composed of 8 plus d. The latitude of motion of a in the first hour is equivalent to d, from the definition of the motion of c. We know that $\dfrac{8+d}{d} = \dfrac{3}{1}$ from supposition 5; hence d is equal to 4. But 4 is the mean of c acquiring latitude from 0 to 8. Q.E.D.[20]

Although Marliani has proved that a latitude of motion uniformly difform from 0 to a certain degree is defined by the true mean between its extremes, he has not yet shown that such is true for a latitude whose initial degree is not zero, i. e. some definite value. For this he poses a body a acquiring latitude uniformly in an hour from 4 to 8 degrees. Then from supposi-tion 6, we know that a moves the same distance as two bodies, one moving with 4 degrees uniformly for the hour, and the other with a uniform degree equivalent to the latitude from 0 to 4, or, from above, to the mean of that latitude which is 2. And a moves then the same distance as one body moving with

---

20 *Ms. cit.*, f. 10r, c. 2, "... patet itaque quod proposita latitudo motus uniformiter difformis a non gradu ad 8 quam credebas non correspondere gradui medio, gradui precise medio eius (in margin) inter extrema, quantum ad spacii pertransitionem correspondet."

a uniform movement composed of 4 plus 2. But the mean of
the latitude acquired by a in the hour is also 6. Q.E.D.[21]

Having proved that a latitude of motion uniformly difform
(a uniformly accelerating velocity) is defined by its mean
degree (velocity), that is for moving an equal space in
an equal time, Marliani now proves that if a body a is moved
uniformly for the first hour with c degree and the second hour
with d degree, it is moved the same distance in these two hours
as a body which is moved uniformly with a degree that is the
mean between c and d. This is accomplished very simply. A
body g moves uniformly with c degrees for one hour; a body
h moves uniformly with d degree for an hour. Then let there
be a body k which is moved uniformly for an hour with a degree of motion composed of c and d; and finally a body e which
is moved with a degree (velocity) that is one half of that of k
for two hours. It is obvious then that e will cover as much
space in two hours as k does in one. From supposition 6, we
know that k covers as much space in an hour as do g and h.
But g and h cover precisely the spaces traversed by a, moving
in the first hour with c degree and in the second with d degree.
Therefore e traveling with one-half the degree of k, i. e. ½
$(c + d)$ for two hours traverses the same distance as a moving under the conditions posited. Q.E.D.

Now Marliani by using only the 6th supposition and its
corollaries and a " very subtle method ", proceeds to prove the
principal conclusion of Dumbleton (Dulmenton), which is our

21 *Ibid.*, f. 10v, c. 1. He puts this proof in another form: ". . . gradus
compositus ex gradu ut 4[or] et ex gradu medio latitudinis (uniformiter
difformis- in margin), a non gradu ad 4[or] distat a non gradu per 3[es]
4[as] (¾) latitudinis, a non gradu ad 8, et gradus vere medius inter
quatuor et 8 distat a non gradu per 3[es] 4[as] latitudinis uniformiter difformis
a non gradu ad 8. Ergo gradus compositus ex 4[or] et ex gradu vere medio
latitudinis uniformiter difformis a non gradu ad quatuor, est vere medium
latitudinis a 4[or] ad 8." (f. 10v, c. 1) That is to say, the value of one part
of the composite degree is 4, or two-fourths of the latitude, a distance 8 from
0 degree, and the value of the other is the mean between 0 and 4, or 2, which
is ¼ of the latitude a distance 8 from 0. Hence the composite degree is the
addition of these two parts, or three-fourths of 8, which is of course the
mean between 4 and 8.

theorem in question. He does this, retaining Dumbleton's method, since it is subtle, and because the copies of Dumbleton's work seem to be corrupt in this passage, at least so it was in his copy and two others he examined.[22]

This passage to which Marliani refers occurs in the *Summa naturalium* of John Dumbleton (ca. 1331-49). In that passage, we see him resort to the indirect method used by Swineshead and Marliani, namely of proving that the unknown degree g can be neither greater nor smaller than the mean degree, and hence must be the mean.[23]

Marliani, then, revises Dumbleton's proof in some detail,[24] the most important part of which may be summarized as follows: A body a acquires latitude uniformly (i. e. accelerates uniformly) in the first hour from o to c degrees, and in the second from c to e at the same uniform rate. We let d be the mean of a's latitude in the first hour. If the latitude of a during this first hour does not correspond to its mean degree d with respect to the amount of space traversed in an equal time, then it corresponds to a degree either more or less than the mean. We first assume that it corresponds to a degree g which is greater than the mean d. In the second hour we know that a acquiring latitude of motion in the second hour from c to e will traverse the same distance as two bodies, in an hour, one moving uniformly with c degree, the other uniformly with g degree. This follows from supposition 6 and its corollaries. Since d is $\frac{1}{2}c$, then a in the second hour will traverse as much space as two bodies, one moving uniformly with d degree for two hours, and the other uniformly for one hour with g degree. Wherefore since a in the first hour will traverse as much space as a body moving uniformly with g degree, it follows that a

<hr>

22 *Ms. cit.*, f. 10v, c. 2. "Non utendo tamen nisi sexta suppositione et corollariis eius possumus secundum medium (!) subtilissimum in hac parte Dulmentonis principalem conclusionem probare, cuius probationem retento eius medio hic ponam quia multum subtilis, et quia libri Dulmentonis in hoc loco adeo corrupti ut quasi nihil accipi possit ex eius dictis, ita erat et liber meus et alii duo quos in hoc passu vidi."

23 Summa naturalium. BN ms. Fonds latin, no. 16140, f. 29v-30r.

24 *Ms. cit.*, ff. 10v-11v.

in those two hours will traverse as much space as will two bodies moving uniformly for two hours, one with d degree of motion (i. e. d velocity), the other with g degree; or consequently, from supposition 6, as one body moving with d plus g degree uniformly for two hours.

If we turn to the convenient method of representing difference in degrees of motion by lines,[25] our proof takes the following direction (See Fig. 2).

25 This method of representing degrees by lines is inferior to Oresme's system of coordinates which has been preserved in two works, *De figuratione potentiarum*, and the *Tractatus de latitudinibus formarum*, the latter being less complete than the former (for a comparison see both Duhem, *op. cit.*, III, pp. 399-400 and H. Weileitner, "Gesetz vom freien Falle in der Scholastik, bei Descartes und Galilei," *Zeitschrift für mathematischen und naturwissenschaftlichen Unterricht*, v. 45 [1914], p. 214). I shall not go into any long summary of Oresme's proof of the theorem in hand, but rather briefly note it (through Duhem, *op. cit.*, III, pp. 388-98). Oresme represents a uniformly difform motion by a triangle, the ordinate representing intensity or degree of motion, the abscissa, time. A constantly uniform motion

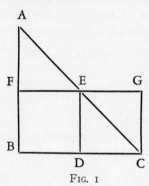

Fig. 1

can by the same coordinates be shown as a rectangle. Hence in this particular problem, Oresme seeks to prove that a uniformly difform motion represented by the triangle ABC, with a latitude AC, whose mean degree is E, is equivalent to the uniform motion of intensity E which has an equivalent abscissa BC (i. e. the time is equal in both cases) and which is represented by the rectangle BFGC. He does so by proving triangles AFE and EGC congruent, and the consequent equality of triangle ABC and rectangle BFGC. (See Fig. 1).

As I have previously indicated, the Italians were aware of Oresme's method. Blasius de Parma (Biagio Pelacani) in his *Questiones super tractatu de latitudinibus formarum* (i. e. questions on the shorter treatise attributed to Oresme) and in the third of his three questions discusses the

We let *b* be the common reference point from which all measurements along the line *be* are taken, while *d, g, c* and *e* are the values already outlined. Then we add to line *bd* the line *bg* and the result is *bh*. It has been proved immediately above that the body *a* in two hours traverses the same space as a body moving uniformly with a velocity of *d* plus *g*; then *h* is the velocity to which the latitude *be* corresponds. From line *bg* Marliani subtracts line *bd*, and there remains *dg*. Similarly from *dh* (equal to *bg*) he subtracts *dc* (equal to *bd*), and

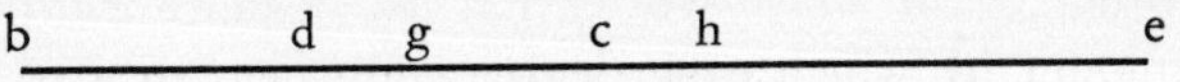

```
b          d    g      c    h                    e
```

FIG. 2

there remains *ch*. Hence since equals subtracted from equals leave equals, *dg* is equal to *ch*. Since the whole latitude *be* corresponds to *h* degree (more intense than *c* the mean degree of *be* by the amount *ch*), and the more remiss half of *be*, i. e. *bc*, corresponds to *g* degree (more intense than its mean degree by the amount *dg*), then therefore it is evident that the degree to which the whole latitude corresponds is more intense than

problem of uniformly difform motion in which we are interested. He affirms that it is equal to its mean, which is proved in a way similar but not as clear as Oresme's, that is by proving the equivalence of a triangle, representing uniformly difform motion, and a rectangle, simple uniform motion. See F. Amodeo. "Appunti su Biagio Pelicani da Parma" in *Atti del IV Congresso internazionale dei Matematici* (Rome, 1908), v. 3, p. 553; also Duhem, *op. cit.*, III, pp. 484-5.

In his commentary on Heytesbury's *Tria predicamenta*, Gaetan of Tiene proves our theorem by use of the familiar triangle and rectangle. The proof is offered after Heytesbury has said that it can be proved that a latitude uniformly difform is equal to its mean (Heytesbury, *De sensu composito*, etc., Venetiis, 1494, f. 40v). "Two is the mean degree of the latitude of the uniformly difform motion. You add up all the degrees above two to the parts less than two, and you have two". (Adde modo omnes gradus supra duo partibus non habentibus duo, et erunt duo). For additional material on the coordinate system consult H. Weileitner, "Ueber den Funktionsbegriff und die graphische Darstellung bei Oresme," *Bibliotheca Mathematica*, 3rd Series, vol. 14 (1913-1914), pp. 193-243 and also by the same author: "Nicolaus Oresme und die graphische Darstellung der Spätscholastik," *Natur und Kultur*, 1916-1917, pp. 529-536; E. Borchert, *Die Lehre von der Bewegung bei Nicolaus Oresme*, Münster, 1934, pp. 92-100 (In *Beitr. z. Gesch. d. Phil.* ..., XXXI, 3).

its mean, by the same amount that the degree to which the more remiss half of the latitude corresponds is than its mean. In the same way, it can be argued that the degree to which bc corresponds is just that much more intense than its mean as is the degree that corresponds to its more remiss half, i. e. bd. This process can be repeated *ad infinitum*. But it is clear that on successive operations of taking the means, soon a particular latitude will be so small that the degree to which that latitude corresponds will be greater than the more intense extreme of that latitude, or in other words that the distance dg, the constant amount by which each degree equivalent to a particular latitude exceeds the mean of that latitude, will soon exceed the distance from the mean degree of a particular latitude to the most intense degree of that latitude, and hence the equivalent degree of that latitude would not be intrinsic. But this conclusion is contrary to the first supposition that the degree to which a latitude corresponds is intrinsic within that latitude. Hence the conclusion of this argument is that the degree to which a latitude is equivalent can not be more intense than its mean degree. This last argument can be shown more clearly by this use of algebraic symbols:

(1) We let $\dfrac{P}{2^n}$ be the value of the nth successive mean of the original latitude, which has an initial velocity of zero and a final velocity of P.

(2) Let V_m be the uniform velocity to which the latitude $O \rightarrow 2\left(\dfrac{P}{2^n}\right)$ corresponds for traversing an equal distance in the same time.

(3) We know that $V_m = \dfrac{P}{2^n} + K$ where K is the constant amount (dg or ch in the argument above) by which the equivalent uniform degree exceeds the mean degree or velocity.

(4) If we increase n toward infinity, $\dfrac{P}{2^n}$ approaches O.

(5) But, Marliani argues, before it becomes O, it must attain some magnitude $\dfrac{P}{2^q}$ which is less than K.

(6) At this point V_m lies without the latitude $O \rightarrow 2\left(\dfrac{P}{2^q}\right)$ since it is equal to $K\left(\text{greater than } \dfrac{P}{2^q}\right)$ plus $\dfrac{P}{2^q}$.

(7) But from the first supposition we know that the uniform velocity to which the change in velocity corresponds for traversing an equal distance in the same time must have some value within that change of velocity (i. e. be intrinsic within the latitude). Hence we conclude that the desired equivalent uniform velocity is not greater than the mean velocity.

Precisely the same method can be used to prove that a latitude can not be represented by a velocity less than its mean. If then the equivalent degree or velocity is neither more nor less than the mean velocity, it must correspond to the mean.[26]

The proof of one or two minor points concludes Marliani's treatise. Perhaps it would be of some value to recapitulate here in a more terse fashion the suppositions and proofs presented in the preceding pages. First the suppositions: (1) That V_m has some value within ΔV, where V_m is the uniform velocity with which one body traverses in a given time the same distance as another moving in the same time with a constantly accelerating velocity ΔV. (2) $\dfrac{a_1}{a_2} = \dfrac{\Delta V_1}{\Delta V_2} = K$. (3) $\dfrac{V_{m_1}}{V_{m_2}} = \dfrac{\Delta V_1}{\Delta V_2} = K$, when $\Delta \dot{V}_1$ and $\Delta \dot{V}_2$ both have an initial velocity of zero. (4) V_m is the same regardless of whether ΔV takes place over a long or short period of time. Marliani deduces as a corollary of supposition four: $\dfrac{\Delta S_1}{\Delta S_2} = \dfrac{\Delta T_1}{\Delta T_2}$ when $\Delta V_1 = \Delta V_2$.

(5) By using these four suppositions, the fifth follows. If a body is uniformly accelerated, it traverses three times as much space in the second half of a given interval of time as in the first, or four times as much space in the total time as in the first half of the interval of time (i. e. if t_2 is twice t_1 then on the

26 *Ms. cit.*, f. 11v, c. 1, "Eodem igitur modo arguendo concluditur quod nulla latitudo uniformiter difformis, non gradu incohata remissiori aut intensiori suo medio gradui correspondet."

basis of $S = kt^2$, $S = 4S_1$). (6) If $V_3 = V_2 + V_1$, then for a given period of time $S_3 = S_2 + S_1$. (A) the first corollary of the sixth supposition states that $S_a = S_b + S_c$ when S_a is the distance traversed by a body a moving with uniform acceleration through a given time from one velocity to another; S_b the distance traversed in the same time by a body b moving with a constant velocity equivalent to the initial velocity of a; S_c the distance described also in the same time by a body c moving with a uniformly accelerated velocity from zero to a velocity equal to the difference in velocities of a. (B) The second corollary holds that $S_a = S_b + S_d$ with the conditions as above except that for S_c is substituted its equal S_d which is the distance traversed by a body moving with a constant velocity having some value within the uniformly accelerated velocity of c.

Marliani's first proof is indirect and is based on showing that certain bodies moving under the conditions necessary to fulfil the corollaries of supposition six would violate those of supposition five unless V_m were the mean of ΔV. The second proof is direct, showing that when the conditions of suppositions five and six are satisfied, then the unknown V_m must equal the mean velocity of ΔV. The revision of Dumbleton's proof, summarized as it has been in the almost preceding pages, should need no recapitulation.

Thus we have examined in detail in this chapter the various proofs given by Marliani of the fundamental theorem of kinematics, that in equal intervals of time a body moving with an uniformly accelerated velocity and one moving with a constant velocity equal to the mean between the initial and final velocities of the first body will traverse equal distances. We have likewise seen that Marliani was not the first to offer a proof of this theorem, that he followed the English school closely, particularly the Calculator, but that his proofs are clearer, if less general, than those of Swineshead. Finally, we should notice the method of composite motions used by Marliani in developing the sixth supposition, and the use he makes of a theory of infinitesimals in revising Dumbleton's proof.

CHAPTER VI

THE PERIPATETIC LAW OF MOTION

It must be made clear at the outset that this chapter does not pretend to be either a complete description of Aristotelian dynamics, nor a detailed study of medieval mechanics. It is only a discussion of a mathematical deduction, or law, if you will, which was a fundamental part of the Peripatetic and late medieval physics. We wish to show particularly how Marliani's *Questio de proportione motuum in velocitate* fits into the history of this law. This chapter will demonstrate that Marliani misunderstood a correction of the law already made in the fourteenth century primarily because he misunderstood the terminology used in that century. As a result, the Milanese physicist has given us only a correction of that early law which might be inferred from the insufficient statements of Aristotle, namely that velocity is proportional to the ratio of the motivating power to the resistance.

Aristotelian dynamics, unaware of any supposition like the Newtonian inertial system, naturally had no conception of force or mass as defined by such a system.[1] It had instead a

---

1 E. Wohlwill has suggested that there were some philosophers in antiquity who may have held a theory akin to the inertial concept. He points to the Peripatetic treatise *Mechanica*, Ch. 8, 35 ff, which declares in answer to the question "Why is it that spherical and circular forms are easier to move?" that "some people say that the circumference of a circle is in movement for the same reason bodies are at rest, because of its resistance (διὰ τὸ ἀντερείδειν). "Die Entdeckung des Beharrungsgesetzes" in *Zeitschrift für Völkerpsychologie und Sprachwissenschaft*, Vol. XIV (1883), p. 370.

Some physicists, particularly Duhem, have seen in the medieval *impetus* theory (which apparently had its origin in the commentary of Johannes Philoponus on Aristotle's *Physics*) the beginnings of the inertial concept. It is my belief, however, that although the impetus theory gave to modern physics a mathematical description of momentum when Jean Buridan described the *impetus* as a function of gravitational mass and velocity (see the long Latin citation of the appropriate passage by E. J. Dijksterhuis, *Val en Worp*, Groningen, 1924, pp. 72-74), it was essentially an invention to

system in which motion originated out of and was maintained
by two opposing powers, motivating force and resistance. To
understand this more completely we must examine briefly the
fundamental points in the Peripatetic explanation. They can
be summarized as follows: (1) Every body has its natural
place at which it would be at rest.[2] (2) There are two types
of local motion deducible from the doctrine of natural place:
natural motion, i. e. when a body by its natural tendency moves

preserve the Peripatetic dynamics rather than to refute it. The dynamics
of Aristotle described motion as arising from the interaction of two forces,
the motivating force and the resistance, and the velocity followed the proportion of the former to the latter. Hence, when it was seen that the *Peripatetic* explanation of the movement of projectiles by the transference of the
motivating power from the projector to the air, which in turn moved and
continued to move the projectile, so that there was always contact between
the motor and the moved, did not fit observation, the philosophers brought
forth the idea of an impressed *impetus*. This *impetus* was said to be impressed
in the moving body by the projector and to move the projectile after there
was not longer contact with the projector. But in such a theory movement
still depended on the interaction of a motivating force (now an impressed
impetus) with a resisting force; while, according to Newton, the motion
imparted to a body by an external force becomes a condition or state of the
body and no interplay of forces is needed to maintain that condition of
motion, which can only be changed by another external force. It is true that
in placing the *impetus* (as an immediate source of motion) within the moved
body a step was made in the right direction. The most important contributions
to the study of the *impetus* theory (none of which is completely satisfactory)
are: E. Wohwill, " Ein Vorgänger Galileis in 6. Jahrhundert " in *Physikal-
ischer Zeitschrift*, Vol. 7 (1906), pp. 23-32; P. Duhem, *Études sur Léonard
de Vinci*, 2nd & 3rd series, Paris, 1909, 1913; B. Jansen, " Olivi der älteste
scholastische Vertreter des heutigen Bewegungsbegriffs " in *Philosophisches
Jahrbuch*, Vol. 33 (1920), pp. 137-152; K. Michalski, " La physique nouvelle
et les differents courants philosophiques au XIVe siècle " in *Bulletin international de l'Academie polonaise des sciences et des lettres*, Classe de philologie. Classe d' histoire et de philosophie, L'Année 1927, part 1 (Janvier-
Mars), pp. 93-164; P. Boutroux, " L'Histoire des principes de la dynamiques
avant Newton" in *Revue de Metaphysique et de Morale*, Vol. 28 (1921),
pp. 657-688; C. Camendzind, *Die antike und moderne Auffassung von Natur-
geschehen mit besonderen Berüchsichtigung der mittelalterlichen Impetus-
theorie*, Lagensalza, 1926; E. J. Dijksterhuis, *Val en Worp*, Groningen, 1924;
M. Deshayes, *Le découverte de l'inertie*, Paris, 1930; K. Hammerle, *Von
Ockham zu Milton*, Innsbruck-Wien-München, 1936; Ernst Borchert, *Die
Lehre von der Bewegung bei Nicolaus Oresme*, Münster, 1934.

2 *De Caelo*, III, ii, 300a-b.

toward its natural position, the heavy things down and the light things up, and the unnatural or violent motion, i. e. when a body is moved away from its natural place by some external power.[3]

(3) All motion whether it is natural or violent depends on the action of two forces: the motivating force and the resistance. In the case of natural motion, such as the fall of heavy bodies, the natural tendency of the body toward its natural place (its weight) becomes the motivating force; the medium provides the resistance. In the case of violent motion, such as that of projectiles, the natural tendency of a body toward its natural place becomes the resistance, while the power that is attempting to draw it from that place is the motivating force.

It is the kind of relationship between these two forces that is of interest to us in this chapter. Although Aristotle's discussion is too brief for an absolute presentation of his views, it has often been supposed that he believed the speed is directly proportional to the motivating force and inversely proportional to the resistance. This is deduced from his remarks concerning both natural and violent movement.

With regard to natural motion he declares in his *De caelo* [4] that " a given weight moves a given distance in a given time; a weight which is as great and more moves the same distance in less time, the times being in inverse proportion to the weights. For instance, if one weight is twice another, it will take half as long over a given movement." In other words, the velocity is directly proportional to the weight.

In another passage the remainder of the proportionality is expressed when the Philosopher tells us that the speed with which a body falls is inversely proportional to the density of the medium: [5] " Now the medium causes a difference because it impedes the body, most of all if it (the medium) is moving in an opposite direction, but in a secondary degree even if it is

3 *Physica*, IV, viii, 215a; VIII, iv, 255b; *De Caelo*, III, ii, 301b.
4 *De caelo*, I, vi, 273b-274.
5 *Physica*, IV, viii, 214a-215b.

at rest, and especially a medium that is not easily divided, a medium that is somewhat dense. A then will move through B in time C, and through D, which is less dense, in time E, the distances traversed through B and D being equal; these movements will follow the ratio of the resisting media . . . Always by so much as the medium is more tenuous, less resisting, and more easily divided, the faster will be the movement."

A longer exposition of Aristotle's basic law of mechanics occurs in the seventh book of the *Physics*. His first observation is that the velocity varies inversely with the resistance. This statement is posed in the form that if A is the moving agent, B the weight of the moved body, C the distance traversed, and D the time taken, then A will move $\frac{1}{2}B$ over the distance $2C$ in time D, or over the distance C in time $\frac{1}{2}D$. He remarks further, in showing that the velocity follows the ratio of the force to the resistance, that if A will move B over a distance C in time D, or over a distance $\frac{1}{2}C$ in time $\frac{1}{2}D$, then E (equal to $\frac{1}{2}A$) will move F (equal to $\frac{1}{2}B$) over C in time D.[6]

According to simple proportionality it would be necessary for E in the example above to be able to move B over $\frac{1}{2}C$ in time D. But no, says Aristotle, such is not the case, for it is possible that E would not able to move B at all. If a divided force could move the total resistance always some definite fraction of the distance which the total force could move that resistance, then it would follow that a single man could haul a ship through a distance the ratio of which to the whole distance it could be drawn by a group of haulers would be equal to the ratio of his individual force to the total force of the group of haulers.[7]

We can not help but wonder why Aristotle did not realize that his own objection vitiated his simple law of proportionality. This objection can be raised in another way. If the motivating force were equivalent to the resistance, Aristotle would no doubt concede, as did the medieval schoolmen, that no move-

6 *Physica*, VII, v, 249b-250a.

7 *Ibid.*, VII, v, 250a.

ment could take place; yet according to the Peripatetic law of proportions, the velocity would have a definite positive value rather than the observed value of zero, or simply, that if $V = K(P/R)$, then when $P = R$, $V = K$ and not zero. As we shall see, Marliani like many of his predecessors was to use this objection to destroy the law of proportionality (See note 41 *infra* and text).

Aristotle did not confine his principle of proportionality to local motion alone, but maintained that it held equally well for qualitative modifications and for growth.[8] The medieval physicists, as we have seen in the earlier chapters, followed Aristotle wholly in applying his law to qualitative changes, especially to heat actions. In fact they went far beyond the Stagirite in the universality of their use of it.

II. The Law in the Middle Ages

The Peripatetic law passed into the West in Latin translations, first of Aristotle's *Physics* (from both Greek and Arabic, ca. 1200), and later of Averroes' commentary thereon (ante 1230?). The internal objection that Aristotle had raised against his own law was of no special concern to the Commentator. He merely included it in rephrasing the law. According to him velocity followed the excess of the power of the motor above the power of the " moved ". To support his statement he notes that by halving the power of the " moved ", the proportion of the motor to the thing moved is doubled and hence the velocity is doubled.[9]

We notice that Averroes, or at least his translator, used the term " excess ". In the middle ages this term was used in two different ways: " arithmetical excess, in which the whole ex-

8 *Ibid.*, VII, v, 250a-b.

9 *Aristotellis ... phisicorum opus cum Averroys ... expositionibus*, Venetiis, Manfredus de Bonellis for Octavianus Scotus, 1495, Bk. VII, f. 121r, c. 2. " ... velocitas propria unicui motui sequitur excessum potentie motoris super potentiam moti et immo cum diversimus motum contingit necessario ut proportio potentie motoris ad motum sit dupla illius proportionis et sic velocitas erit dupla ad illam velocitatem."

cedes the part, or the greater excedes the lesser, and geometrical excess, which is the proportion of the greater to the lesser." [10] The schoolmen rightfully assumed that Averroes was speaking of " excess " in the latter, geometrical sense, since the example he cites of the effect of halving the power of the " moved " can lead to no other interpretation of the word.

The thirteenth century commentators followed the Aristotelian tradition without dissent, accepting the Commentator's incorporation of Aristotle's objection into the law. However they did not limit themselves to the term " excessus " to express the idea that motion could take place only when the power of the motor (the first term of the proportion) was greater than the power of the thing moved (the second term of the proportion). " Victoria " and " dominium " were also used.[11]

Parallel with the rise of kinematics in the fourteenth century, some of the results of which we have examined in the preceding chapter, there was a corresponding intensification of interest in dynamics. Special treatises were composed to investigate the Aristotelian rules for comparing movements. One of the first of these treatises was that of Thomas Bradwardine,[12] procurator

---

10 Paul of Venice makes this distinction in his *Summa Naturalium*, Milan, Christopher Valdaser, 1476, quest. 32 on the *Physics* (no pagination). "... duplex est excessus, scilicet arismetricus quo totum excedit partem, vel maius excedit minus, et geometricus, qui est proportio maioris ad minus, de primo excessu non loquebatur commentator, sed de secundo." *Cf.* W. Burley, *In octo volumina ... Aristotelis de physico auditu expositio*, Padua, 1476, ff. 233r-234v.

11 Thomas Aquinas, *Opera Omnia*, vol. 2, *Commentaria in octo libros physicorum Aristotelis*, Rome, 1884, pp.. 358-359. *Cf.* Albertus Magnus, *Opera omnia* (Ed. Borgnet), vol. 3, Paris, 1890, Lib. VII, Tract. II, Chap. V, pp. 516-517, and Egidius Romanus, *Commentaria in VIII libros physicorum*, Patavii, Hyeronimus Durantis, 1493, Lectio X (no pagination).

12 It seems to me quite possible that the fragment cited by Duhem as Ms. BN Fonds latin, 8680 A, ff. 6-7, which appears to have been noted by Bradwardine as *De proportionalitate motuum et magnitudinum* is only a part of a longer treatise which like Bradwardine's included a discussion of the Peripatetic law. See Duhem, *Études*, III, p. 292. Duhem's citation is in error. Dr. Moody of Columbia University sent for the folios indicated and found

of the University of Oxford in 1325, Archbishop of Canterbury in 1349, and dead in the same year. This work, known
as the *Tractatus proportionum,* was composed in 1328.[13] It
consists of an extensive treatment of the Peripatetic law and
some kinematic problems. We are interested in this chapter only
in Bradwardine's discussion of the law. It was his discussion
that became the standard for the next two centuries and which
Marliani attempted to refute in presenting his " correction ".

Marliani objected to the law of simple proportionality because of an inconsistency therein. He claimed that a velocity
resulting from a quadruple proportion should be, according
to the law, greater (actually four times greater) than that
resulting from a proportion of equality (i. e. where the
motivating force is equal to the resistance). But all the
physicists recognized that no motion could result from a proportion of equality.[14] The two considerations are mutually inconsistent. Therefore the law must be amended to fit the observed fact that no movement takes place when the force equals
the resistance.

Although Marliani accuses Bradwardine of falling into the
same error, actually the Englishman does not. The purpose of
his preliminary treatment is to show that comparisons can
only be made between proportions of the same kind, i. e. between one proportion of greater inequality and another, or
between one proportion of less inequality and another, but
never between a proportion of greater inequality and a pro-

them to be from a work of Jordanus on weights. I might take this opportunity to mention Dr. Moody's help in interpreting Bradwardine's work.

13 Three manuscripts at the Bibliothèque Nationale in Paris give this date:
Mss. no. 16621 (f. 212v) ; 14576 (f. 261) ; nouv. acquis. lat. no. 625 (70v).
The first two have been cited by Duhem, *Études,* III, p. 299, and all three by
Thorndike, *A History of Magic and Experimental Science,* III (1934), p. 376.

14 In the ensuing discussion we have retained the terminology of the
medieval treatises on proportions. A proportion of equality may be defined as
$\frac{a}{b} = 1$.   A proportion of greater inequality is $\frac{a}{b} > 1$ and a proportion of
lesser inequality, $\frac{a}{b} < 1$.

portion of equality, or between a proportion of greater inequality and one of lesser inequality. The object as he states
it is threefold: (1) no proportion of equality can be more or
less than another proportion of equality; (2) no proportion of
greater inequality can be more or less than a proportion of
equality; (3) no proportion of greater inequality can be more
or less than a proportion of lesser inequality.[15] Since movement was thought to rise only from proportions of greater
inequality, it is quite obvious why Bradwardine has carefully
limited comparisons of proportions of greater inequality to
other proportions of greater inequality. On the basis of simple
proportionality these statements would be incorrect. But Bradwardine is referring them to the proportion of proportions.
Just as there is no proportion in velocity between rest and
motion, in the same way there is no proportion of greater inequality between the unequal and the equal. He is thinking in
terms of a geometric progression of the original proportion
thus: a/b, $(a/b)^2$, $(a/b)^3$, . . .

Bradwardine's first conclusion relative to the theory of proportions tells us that in the case of three continually proportional terms, if the proportion of the first to the second and
the proportion of the second to the third are both proportions
of greater inequality, then the proportion of the first to the
third is precisely double *(dupla)* that of the first to the second.[16]
In the case of four terms, the first to the fourth would be triple
the first to the second, etc. *Dupla, tripla,* etc., as used here by
Bradwardine, do not mean " twice ", " thrice ", etc., as Marliani
would have us believe in attacking the Englishman, but rather

15 Bradwardine, *Tractatus proportionum* (ed., Venetiis, Bonetus Locatellus,
1505, ff. 11v-12r; Ms. BN (Paris) Fonds latin, 6559, f. 51r-v). " Nam nulla
proportio equalitatis alia proportione equalitatis est maior vel minor . . . Nulla
ergo proportio maioris inequalitatis proportione equalitatis est maior nec
minor . . . Nulla proportio maioris inequalitatis alia proportione inequalitatis
minoris est maior vel minor . . . "

16 *Ibid.*, Ms., f. 51r; Ed., f. 11v, c. 1. " Si fuerit proportio maioris inequalitatis primi ad secundum ut secundi ad tertium, erit proportio primi ad tertium precise dupla ad proportionem primi ad secundum."

" squared ", " cubed ", etc. His conclusions and examples bear this out. The point that confused Marliani was that Bradwardine used *dupla* in both senses of the word—sometimes in the same conclusion! When used for individual numbers or for the value of an original proportion, it would mean " twice "; when used in comparing proportions, its signification would be " squared." Such usage was common with Oresme, Dumbleton, and Albert of Saxony.

Bradwardine's first conclusion can be found in the tenth definition of the *Elements* of Euclid. Euclid declares that when three quantities are continually proportional, the proportion of the first to the third is equivalent to the proportion of the first to the second duplicated.[17] Campanus correctly adds, in commenting on the definition, that Euclid means that the proportion of the first to the third is composed of two such proportions of the first to the second, i. e. of the proportion of the first to the second multiplied into itself *(in se multiplicata)*.[18]

Believing erroneously that Bradwardine has misunderstood Euclid and Campanus, Marliani hotly attacks Bradwardine's conclusions. He points out that the value of the proportion, which is spoken of as its denomination *(denominatio)*, is the only means by which one proportion can be compared to another. A proportion which is equivalent to another proportion duplicated can never be also twice that proportion except when that proportion has a value of two. A proportion which is equivalent to another proportion triplicated (cubed) can never be under any circumstances triple that proportion etc. Marliani makes constant use of Jordanus, Alkindi, Campanus, and Euclid to support his position.[19]

17 Euclid, *Elementa* (with commentaries of Campanus), Venetiis, Erhardus, Ratdolt, 1482 (no pagination), lib. V, def. 10. " Si fuerint tres quantitates continue proportionales dicetur proportio prime ad tertiam proportio prime ad secundam duplicata. (I have used an early Latin translation in order to illustrate the terminology with which the medieval physicists were familiar.)

18 *Ibid., loc. cit.*

19 *De proportione motuum in velocitate*, Pavia, D. de Confaloneriis, 1482, ff. 10v-12r.

Marliani believes that the reason why Bradwardine has confused *dupla* and *duplicata* is that he does not understand correctly the " composition " of proportions. " Composition " with
respect to proportions of course means the *multiplication* of
one of the component proportions by the other (or others),
and not the *addition* of those proportions, as the Englishman's
conclusions lead Marliani to believe. For instance, Bradwardine's third conclusion declares that if the first term is
more than double the second, and the second is exactly double
the third, then the proportion of the first to the third is less
than twice (double) the proportion of the first to the second.[20]
Marliani thinks that Bradwardine is trying to prove $a/c < 2a/b$
when $a/b > 2$ and $b/c = 2$. The Milanese physician is led to the
conclusion that this can be proved only by understanding the
composition of proportions in the sense of the addition of proportions (see text following note 27 for a false proof of this).
In reality, Bradwardine wishes to prove that if $a/b > 2$ and
$b/c = 2$, then $a/c < (a/b)^2$, which is demonstrable on the basis
of the true conception of composition of proportions. Marliani
also wrongly believes that Bradwardine's fourth, fifth, and sixth
conclusions can be proved only on this mistaken idea of the
composition of proportions.

Marliani's misunderstanding of Bradwardine's conclusions
on proportions naturally leads him to a similar misunderstanding of the Oxford schoolman's conclusions on velocity.[21]

Having presented these initial observations on the theory
of proportions in order to maintain the consistency of the law,
Bradwardine next rejects four erroneous opinions as to the
proportionality that velocity follows, and then finally offers the
fifth one, which he believes to be correct. These opinions are
rejected by Bradwardine (and also either partly or fully by
many others, including Albert of Saxony, Nicholas Oresme,

---

20 *Tractatus proportionum*, Ms., f. 51r; Ed., f. 11v, c. 2. " Si fuerit primum
maius quam duplum secundi, fueritque secundum equaliter duplum tertii, erit
proportio primi ad tertium minor quam dupla ad proportioem primi ad
secundum.

21 See *Questio de proportione*, ff. 16-17.

Auctor de sex inconvenientibus, John of Dumbleton, Paul of Venice, etc.[22] They describe velocity following: (1) the excess of the power of the motor to the power of the thing moved; (2) the proportion of the excess of the power of the motor to the power of the thing moved;[23] (3) the proportion of the passives (i. e. the things moved) with the motor remaining the same, or the proportion of the motors with the passives remaining the same; and (4) no proportion or excess at all, but a *dominium* or natural habitude *(habitudo)* of the motor to the thing moved.[24]

After deciding against these four theories, Bradwardine advances the common theory that the proportion of velocities follows the proportion of the power of the motor to the power of the thing moved.[25] This expression of the Peripatetic law received almost universal acceptance in the fourteenth and fifteenth centuries.[26] Marliani confuses this representation of

22 Albert of Saxony, *Tractatus de proportionibus* (with Walter of Burley, *De intensione et remissione*, Venice, 1496) f. 44r, c. 1. Nicholas Oresme, *Tractatus proportionum* (with B. Politus, *Questio de modalibus*, Venice, 1505), f. 17r, c. 1. See also *Tractatus de sex inconvenientibus* (with B. Politus, *op. cit.*, Ed. cit.), f. 49r, c. 1. See also Ms. Paris BN Fonds latin, 6559, f. 28v, c. 1; John of Dumbleton, *Summa naturalium*, Ms. Paris, BN Fonds latin, 16146, ff. 27r-28r; Paul of Venice, *Summa naturalium*, Milan, C. Valdarser, 1476, question 32 on the *Physics* (no pagination).

23 The first of these seems to mean that velocity follows simple arithmetical excess, while the second is possibly the theory later adopted by Marliani, that velocity follows the proportion of the arithmetical excess of power of the motor above that of the power of the thing moved to the power of the thing moved. In his refutation of the third of these erroneous laws, Bradwardine advances the same argument that Marliani was to employ, namely that no motion can arise from a proportion of equality.

24 *Tractatus proportionum*, Ms., ff. 51-54; Ed., ff. 12r-13v.

25 *Ibid.*, Ms., f. 54v; Ed., f. 14r, c. 2. "... proportio velocitatum in motibus sequitur proportionem potentie motoris ad potentiam rei moti..."

26 See the various citations in note 22 *supra* for the discussions of the law by Albert of Saxony, Nicholas Oresme, Auctor de sex inconvenientibus, John of Dumbleton, and Paul of Venice. Consult also the first section of the fourth part of the *Quadripartitum numerorum* of Jean de Murs, entitled *De moventibus et motis*, which discusses and accepts the law (Paris, BN Fonds latin, 7190, ff. 72-81. *Cf.* Duhem, *Études*, III, pp. 300-301). Marliani numbers Buridan among those who accepted the common opinion of Brad-

the law with the third erroneous opinion already rejected by Bradwardine. He makes this mistake by misunderstanding what was meant by " proportion of proportions." Marliani continued to compare the proportions arithmetically, so that the proportion of velocities reduced to velocity following a simple ratio of motor to moved. However, when Bradwardine and his followers spoke of the proportion of velocities following the proportion of proportions, they were thinking in terms of the original proportion squared, cubed, quadrupled, etc. (or, in other words, in terms of a geometric progression).

Following his general statement of the law, Bradwardine proceeds to some additional conclusions.

The Englishman first claims that if the proportion of the motive power to the resistance is a double one $\left(\text{i. e. } \frac{b}{c} = 2 \right)$, then when the motive power is doubled, it will move the same resistance twice as fast.[27] This conclusion is given in order to " save " Aristotle by showing that $(b/c)^2 = 2 \times 2/1$ or $2/1 + 2/1$. Marliani asserts that Bradwardine derives this conclusion not from the relationship $\frac{a}{c} = \frac{a}{b} \cdot \frac{b}{c}$ which would be the correct interpretation of the composition of proportions, but rather from the incorrect $\frac{a}{c} = \frac{a}{b} + \frac{b}{c}$, which happens in the case where $\frac{a}{b} = \frac{b}{c} = 2$ to provide the same answer. This can be seen by assuming that velocity $\frac{b}{c} = 2$ and that when $\frac{a}{b} = \frac{b}{c}$ then $a = 2b$. Therefore $\frac{a}{c}$ is precisely double $\frac{b}{c}$.

The rest of Bradwardine's conclusions however do not offer, Marliani believes, the same coincidence as to the methods of

wardine, along with Albert of Saxony, Oresme, Marsilius of Inghen, Paul of Venice, and "quasi omnium modernorum" (*De proportione*, Ed. cit., f. 12r, c. 2). Walter of Burley, *In octo volumina ... Aristotelis de physico auditu expositio*, Padua, 1476, f. 232v, c. 2 and Richard Swineshead, *Liber calculationum*, Padua, ca. 1477, f. 52v, c. 1) follow Bradwardine.

27 *Op. cit.*, Ms., f. 54v; Ed., f. 14v, c. 1. " Si (proportio) potentie moventis ad potentiam sui moti sit dupla proportio, potentia motiva duplicata movebit idem motum precise in duplo velocius."

" composing " proportions. He believes their proof rests on the false relationship $\frac{a}{c} = \frac{a}{b} + \frac{b}{c}$. Let us take as example the fourth conclusion which states that if the proportion of the motive power to the power of the thing moved is greater than a double proportion, then the velocity with which double the motive power moves the same resistance is never doubled.[28] This conclusion is absolutely inconsistent with the law of direct proportionality when the true method of the composition of proportions is employed, but on the basis of $\frac{a}{c} = \frac{a}{b} + \frac{b}{c}$ it can be proved as follows. If $\frac{b}{c}$ is greater than 2 and a is double b, then $\frac{a}{c}$ is never double $\frac{b}{c}$. Translating into algebraic terms, the proof becomes:

To prove:

(1) $\dfrac{\frac{a}{c}}{\frac{b}{c}} < 2$ when $\frac{a}{c} = \frac{a}{b} + \frac{b}{c}$ and $\frac{a}{b} = 2$, $\frac{b}{c} = 2 + e$,

(2) $\frac{a}{c} = 4 + e$, $\dfrac{\frac{a}{c}}{\frac{b}{c}} = \frac{4+e}{2+e} = 1 + \frac{2}{2+e}$.

(3) Since $\frac{2}{2+e} < 1$, then $\dfrac{\frac{a}{c}}{\frac{b}{c}} < 2$. Q. E. D.

Bradwardine, however, does not accept the law of simple proportionality. He is actually trying to prove $a/c < (b/c)^2$, which can be easily proved using the correct idea of the composition of proportions.

We have outlined up to this point the early development of the law of proportionality and have studied in some detail how it received its most usual correction at the hands of Thomas

28 *Ibid.*, Ms., f. 54v; Ed., 14v, c. 1.

Bradwardine. We have in addition pointed out that it was almost universally accepted in the middle ages in the form enunciated by Bradwardine.[29] Now it remains to examine Marliani's discussion of the law.

From almost the beginning of his career the Milanese physician was suspicious of the law. He tells us in his *De proportione* that he had always been very much nonplussed when lecturing publicly on the *Proportiones* of Thomas Bradwardine and Albert of Saxony, for these men seemed to demonstrate their conclusions with such weak and inconclusive reasoning. Marliani further adds that later he wrote in his first *Tractatus de reactione* that he had spoken of the proportion of motions according to the common opinion in that work, but that in the future he would try to find a new or different opinion to follow.[30] However, he continued to use the Aristotelian law somewhat later in his discussions with Giovanni Arcolani.

In his *Liber conclusionum diversarum* Marliani raised certain " difficulties " against the law. He speaks of the *difficultates* sent to Philip Adiuta (See Chapter I, section G) as being conclusions 58-61 of the above work.[31] These *difficultates* together with answers to them, written after the *Liber conclusionum diversarum* and before the *Questio de proportione,* pose two types of problems which seem to be incompatible with the com-

---

29 It is quite probable that there were a number of people of whose names we are ignorant who did not accept the law, since we have seen that Bradwardine and others always devoted some attention to other opinions.

30 *De proportione* (1482), f. 15v, " Et fateor me semper plurimum admirationis accepisse cum legerem publice proportiones Braduardini aut Albertoli, quoniam scilicet ratione tam debili nihilque concludenti conclusiones suas demonstrare videbantur, Et proinde scripsi in tractatu de reactione primo me in illo de proportione motuum secundum communem opinionem (dixisse), et quod [in] futurum novam aut saltem aliam a communi aperirem viam." The passage of the *De reactione* to which Marliani refers occurs in the third part (Ms. San. Marco VI, 105, f. 40r, c. 2) : " Et quamquam in materia de proportione motuum nunc credam innovare rationes, non tamen ille viciabunt que hic dicta sunt, sed solum aliqualiter erunt secundum illas moderanda."

31 *De proportione,* f. 3v, c. 1.

mon law. The first of these professes to prove that it is possible for unequal powers encountering similar resistances to produce equal velocities. The second shows how equal powers can produce equal velocities even against unequal resistances.[32] Difficulties of the same nature are included in the first section of the *Questio de proportione*.[33] When elaborating on the difficulty of accepting the law in his dispute with Philip Adiuta, Marliani remarks further that he has written elsewhere and publicly spoken against the common view.[34]

Although he had questioned the Peripatetic law in these earlier works, the Pavian doctor did not present a well developed theory of his own until the composition of his *Questio de proportione motuum in velocitate* in 1464 (See Chapter I, section I). This treatise, like the *Questio de caliditate,* is strictly scholastic in form. The first part includes the common arguments brought against the law, many of which Marliani recognized as erroneous. The second part discusses the mathematical theory of proportions and is mainly a refutation of Bradwardine's conclusions. The third presents Marliani's own conclusions including certain objections to those conclusions together with their solution. The concluding section evaluates the arguments brought against the law in the first part.

Marliani cites a wide variety of sources in this treatise. Among these sources are the mathematical works of Euclid, Alkindi, Jordanus, and Campanus; the *Proportiones* of Bradwardine, Albert of Saxony, and Nicholas Oresme; the commentaries on Aristotle's *Physics* of Jean Buridan and Marsilius of Inghen; the commentaries *Summa naturalium* of John of

---

32 *Difficultates misse per J. de Marliano Philippo Adiute Veneto* (published in the second volume of Marliani's collected works, Pavia, ca. 1482. See Chapter I, section M). The first part (ff. 69v-71r) presents the "difficulties". The remaining parts include responses by Philip and elaborations by Marliani (ff. 71r-80v).

33 Ed. cit., f. 2v, ff.

34 *Difficultates*, Ed. cit., f. 77v, "Licet de proportione motuum an scilicet talis si qualis proportio proportionum multa alibi scripserim et publice tenerim contra communem usque nunc opinionem...."

Dumbleton and Paul of Venice; the *Logica* and *Metaphysica* of Wyclif; the *Sophismata* of Richard Clienton; the *Tria predicamenta de motu* of William Heytesbury; a treatise on motion of Messinus (possibly his commentary on Heytesbury's *De motu*); and the *Liber calculationum* of Richard Swineshead (cited by Marliani as the Calculator).

Most of the arguments in the first part of the *Questio de proportione* are of no interest to us in this chapter, since, as Marliani at least partially recognizes, they are based on faulty reasoning. However, it can be noted that a few of these, in which the reasoning is bad, have some value because they are based on experimentation. One involves the use of the pendulum *(festuca supra pondus)* [35] and another of spherical weights upon inclined planes.[36]

We have already cited passages from the second part which illustrate Marliani's opposition to Bradwardine. This opposition is further expressed in Marliani's conclusions. The first conclusion affirms against Bradwardine that any proportion of greater inequality is greater than a proportion of equality and that any proportion of greater inequality is greater than a proportion of lesser inequality.[37]

Marliani's second conclusion takes issue with the first of Bradwardine. It declares that if there are three terms continually proportional, the proportion of the first to the third will not be truly and precisely double *(dupla)* the proportion of the first to the second although it is said to be equivalent to that proportion duplicated *(duplicata)*.[38] This leads us to a definition of "double". A proportion whose denomination (value) is twice that of another proportion is said to be double

35 *De proportione*, f. 4r, c. 1.

36 *Ibid.*, f. 4r, c. 2.

37 *Ibid.*, f. 16r, c. 1.

38 *Ibid.*, f. 16r, c. 2. The third and fourth conclusions repeat this conclusion for four and five proportional terms, substituting "triple" and "triplicated", "quadruple" and "quadruplicated" for "double" and "duplicated" (f. 16v, c. 1).

it.[39] A proportion that is known of as a " double proportion "
(i. e. one with a denomination or value of 2) is double a proportion of equality, while a " subdouble proportion " (with a
denomination of $\frac{1}{2}$) is one-half of a proportion of equality.[40]

Since these aforementioned conclusions are true, it is quite
obvious that velocity or the " proportion of motions in velocity " does not follow the proportion of the motive powers
to the resistances. For a quadruple proportion (one whose denomination is 4) is greater (actually four times greater) than
a proportion of equality (whose denomination is of course 1).
Yet while there is a certain velocity arising from a quadruple
proportion, there is, according to the common opinion, no
velocity or motion arising from a proportion of equality.[41] The
inconsistency is clear.

Since this inconsistency makes untenable the supposition
that velocity follows the ratio of the power of the motor to
that of the thing moved, what theory can be advanced which
will not suffer the same criticism? Marliani answers that the
inconsistency is no longer apparent if velocity follows the proportion of the excess of the motive power above the resistance
to the resistance $\left(V = K \dfrac{P - R}{R} \right).$[42] This is true because
when the power is equal to the resistance, the velocity will
equal zero.

39 *Ibid.*, f. 16v, c. 1. " Proportio cuius denominatio dupla est (ad) denominationem alicuius proportionis dupla est ad illam."

40 *Ibid.*, f. 16v, c. 2.

41 *Ibid.*, f. 17r, c. 1. " ... proportio quadrupla est maior proportione
equalitatis. Ergo non qualis est proportio proportionum talis proportio velocitatum ... consequentia est manifesta, nam a proportione quadrupla sit
motus certe velocitatis; a proportione vero equalitatis nullus sit motus ... "
Marliani remarks further that Wyclif, Dumbleton, and Messinus (the unknown commentator on Heytesbury's *De motu*) recognized this objection
and failed to solve it satisfactorily.

42 *Ibid.*, f. 18v, c. 1. " Universaliter talis est proportio motuum mere
naturalium et vere sucessivorum atque eiusdem speciei qualis est proportio
proportionum excessuum potentiarum motivarum supra suas resistentias ad
suas resistentias." Marliani interprets proportion of proportions not in the
geometric sense conceived by the fourteenth century schoolmen, but rather
as a simple arithmetical comparison of one proportion to another.

Marliani cites in support of this theory the statement of Averroes noted above that velocity follows the excess of the motive power above the power of the thing moved.[43] There can be no doubt that Marliani has misunderstood Averroes, for the Commentator's statement that doubling the proportion of the power of the motor to that of the thing moved produces a doubled velocity indicates he follows the common understanding of excess as the proportion of the greater to the lesser rather than the difference of the greater and the lesser.

The Milanese physician wisely points out that the example of the ship haulers that Aristotle cited to show that divided power need not necessarily move the total resistance is in support of his theory. Yet Marliani was unable to abandon the Stagirite totally even when it was clear that their theories were in conflict. As a result Marliani's ninth and tenth conclusions are partially inconsistent with his own law. The ninth declares that if any force moves some body resistant to it with a certain velocity, every force double that first force and applied to a body of the same resistance will move that body twice as fast or more than twice as fast as the first force.[44] This conclusion is partially incorrect according to Marliani's own law, because when the power is doubled and the resistance is constant, the velocity is never precisely doubled, but is always more than doubled. Algebraicly this criticism can be shown as follows:

If $V_1 = k \dfrac{P_1 - R_1}{R_1}$ when $P_2 = 2P_1$ and $R_2 = R_1$ then

$V_2 = k \dfrac{2P_1 - R_1}{R_1} = k \dfrac{2P_1}{R_1} - k$ which obviously exceeds $2V_1$

which equals $k \dfrac{2P_1}{R_1} - 2k$.

It is possible that this criticism is unjust and that Marliani was merely presenting us with two alternatives, but wished us

43 See notes 8 and 9 *supra* with the text.

44 *De proportione*, f. 19r, c. 1. " Si aliqua potentia moveat aliquod mobile sibi resistens aliqua velocitate, omnis potentia dupla ad illam mobili equaliter cum primo resistenti applicata et per tempus movebit suum tale mobile in duplo velocius aut plus quam in duplo velocius quam dicta potentia."

to consider as true only the last part of the conclusion. For his proof, which is an indirect one, is based on showing that if this conclusion were not true, then there would be some force double the first force which would not move the same resistance *more than twice as fast* as the first moved it.[45] If Marliani were trying to prove the whole conclusion, i. e. a doubled force would move the same resistance either twice as fast or more than twice as the original force, the indirect proof should have read: If this conclusion were not true, then there would be some force double to the first force which would not move the same resistance *either twice as fast or more than twice as fast* as the first moved it.

Another point that lends credence to the possibility that Marliani did not actually believe that doubling the power ever resulted, according to his formula, in doubling the velocity, is that he clearly recognizes in the sixteenth conclusion that the opposite operation of taking one-half of the power would not produce a velocity one-half as fast.[46]

The tenth conclusion is subject to the same criticism as the ninth. It involves a non-variable force which acts on half the original resistance and thereby produces a velocity which is twice as great or more than twice as great.[47]

The next two conclusions are worded so as to contrast with conclusions of Bradwardine. We are told that in no given proportion which is greater than two will the power " duplicated " move the same resistance precisely twice as fast,[48] nor will the same power move half the resistance twice as fast.[49] The last part of this statement seems to bear out our suspicion that, like the ninth, the tenth conclusion should perhaps be considered as presenting two alternatives, the second of which Marliani actually believed.

45 *Ibid.*, f. 19r, c. 1. " Hec conclusio patet; nam si non est ita detur potentia aliqua que dupla ad primam non moveat equale mobile plus quam in duplo velocius quam prima suum."

46 *Ibid.*, f. 20v, c. 1. " Non sit aliqua potentia movet aliquod mobile aliqua velocitate, medietas potentie movebit idem mobile in duplo tardius."

47 *Ibid.*, f. 19r, c. 2.   48 *Ibid.*, f. 19r, c. 2.

49 *Ibid.*, f. 20r, c. 1.

Marliani's remaining twelve conclusions describe how the velocity is altered as different variations are made in the force and resistance. They include cases where half the force is taken at the same time as half the resistance, two forces that separately move their resistances equally fast are joined together to move their resistances joined together, and two forces that separately move their resistances unequally fast are joined together to move their resistances joined together.[50] Likewise these conclusions treat more generally of instances when uniform variations take place in force and resistance.[51]

Marliani terminates his conclusions with the remark that many of the conclusions that the Calculator had developed in his *Tractatus de regulis motus localis* (the last part of the *Liber calculationum*) would result just as well from Marliani's law as the common one.[52]

The Milanese physician apparently had little success in substituting his theory in place of Bradwardine's law. Two later physicists, Alessandro Achillini[53] and Benedetto Vittori,[54] reject his explanation in favor of the latter. Leonardo da Vinci also was familiar with the *Questio de proportione* of Marliani, but although he criticizes Albert of Saxony's exposition of the Peripatetic law in one place, there is no evidence that this criticism follows that of Marliani.[55]

50 *Ibid.*, f. 20r-v.

51 *Ibid.*, ff. 20-21.

52 *Ibid.*, f. 21r, c. 2.

53 *De proportione motuum* in his *Opera omnia*, Venice, 1545, f. 192v, c. 1.

54 *Commentaria in tractatum proportionum Alberti de Saxonia*, Bononie, Benedictus Hector, 1506. There is a long exposition of Marliani's theory in this work, but unfortunately after noting briefly that it was against Marliani, I put it aside for later examination. The war, however, closed the Bibliothèque Nationale before I had returned to it.

55 In the *Codico atlantico* Leonardo speaks of the *Proportions* of Alchino with the considerations of Marliani (Duhem, *Études*, III, p. 511). Also Ms. F of the Bibliothèque de l'Institut in Paris (published by Charles Ravaisson-Mollien, Paris, 1889) has a notation on the verso of the cover which seems to be to the *De proportione*: "Albertuccio et Marliano decalulatione".

CHAPTER VII

VULGAR FRACTIONS

We now turn our attention from Marliani's physical theories to his only extant mathematical work, the *Algorismus de minutiis* (see Chapter I, Section H).

The first part of this chapter will be devoted to a short chronological list of a number of treatises on vulgar or ordinary fractions from the time of their origin and growth among the Indians through their later development by the Arabic-speaking, Jewish, and Western peoples. The second section will be a comparison of these various works with that of Marliani.

Of the Indian treatises, where the operations with common fractions possibly originated, there are five in particular that can serve as a basis for a study of Indian fractions.[1] The first of these is the twelfth chapter of the astronomical work, the *Brahma-sphuta-siddhanta* of Brahmagupta (7th c.),[2] which was followed in the next century by the *Trisatika* of Sridhara (ca. 750),[3] and in the ninth century by the *Ganiat-sara-sam-graha* of Mahavira (ca. 850).[4] We can cite finally two other

1 The most extensive treatment of Indian mathematics as a whole is that of B. Datta and A. N. Singh, *History of Hindu Mathematics,* Part I, "Numerical Notation and Arithmetic," Lahore, 1935. It is particularly useful in that it is a source book including a great number of passages not hitherto translated. The authors often question earlier translations, such as those of G. R. Kaye; while they pay tribute to the editions of P. S. Dvivedi of Benares. Gandz, however, claims that this work is not too reliable, especially for early dates, *Isis,* Vol. XXV (1936), pp. 478-88.

2 Edited by Dvivedi (P. S.), Benares, 1902, and translated into English by H. T. Colebrooke, *Algebra with Arithmetic and Mensuration from the Sanscrit of Brahmagupta and Bhascara,* London, 1817.

3 Edition by Dvivedi, Benares, 1899. Translated in part by G. R. Kaye and N. Ramanujachariar in *Bibliotheca Mathematica,* series III, vol. XIII (1912-13), pp. 203-17.

4 Edited and translated by M. Rangacarya, Madras, 1912.

Indian sources of importance, the *Mahasiddhanta* of Aryabhata II (ca. 950) [5] and the *Lilavati* of Bhaskara II (1150).[6]

The Arabic discussions, which combine the Indian methods of dealing with ordinary fractions, first with the Greek-Egyptian usage of unit fraction, and then with the sexagesimal treatments derived from Babylonian sources, are numerous, but there are six which seem to be the most important. We may begin with a treatise that is attributed to al-Khwarizmi (fl. 813-33), but which remains today only in a Latin translation of the twelfth century, the *Algoritmi de numero Indorum*.[7] There is some reason to believe that this treatise which contains only slight reference to ordinary fractions, but a more extended study of sexagesimal fractions, is incomplete, and that the part omitted may contain a longer summary of vulgar fractions.[8] The second of the Islamic arithmetics is that of al-Nasawi (before 1030).[9] In the same century we have the less important treatment of al-Karkhi (ca. 1010-6),[10] and a century later the complicated work of al-Hassar,[11] followed by its derivative, the thirteenth century *Talkis* of Ibn al-Banna, which possesses a much improved summary of the operations with fractions.[12]

5 Edited by Dvivedi, Benares, 1910.

6 Edited by Dvivedi, Benares, 1910; translated into English by H. T. Colebrooke, *op. cit.* in note 2 *supra*, London, 1817.

7 Published by B. Boncompagni, *Trattati d'aritmetica*, I, Rome, 1857.

8 J. Tropfke, *Geschichte der Elementar-Mathematik*, 3rd edition, vol. I, Berlin, 1930, p. 160.

9 The introduction is translated and the contents incompletely analyzed by F. Woepcke in *Journal Asiatique*, 6th series, vol. I (1863), pp. 491 to 500. It is further analyzed by H. Suter in *Bibliotheca Mathematica*, 3rd series, vol. VII (1906), pp. 113-119.

10 Translated into German in three parts by Ad. Hochheim, Halle, 1878-80. Hindu numbers are not used in the text, and it seems to be based on Greek and Hellenistic sources.

11 Described by H. Suter, *Bibliotheca Mathematica*, 3rd series, vol. II (1901), pp. 12-40.

12 Translated into French by Aristide Marre, *Atti dell' Accad. Pont. de Nuovi Lincei*, vol. 17 (1865), pp. 289-319.

The final medieval Islamic treatise is that of al-Kalcadi (d. 1486), known as *Lifting the Veil from the Gobar Science*.[13]

Of the Jewish authors, who owed their knowledge largely to their Arabic-speaking predecessors, we shall employ only one, Abraham Ibn Ezra (1092?-1167), who composed the *Sefer ha-mispar*.[14]

The Western authors in theory and originality do not go beyond the Hindu treatises. They have their immediate origin in Arabic sources, although they are somewhat influenced by earlier arithmetics based on Roman sources, such as the *De minutiis* of Demetrius Alabaldus (d. 1026).[15]

We turn first to the twelfth century, the century of extensive translations into Latin from the Arabic. It yields not only the translation of the work attributed to al-Khwarizmi already mentioned (the *Algoritmi de numero Indorum*), but also at least two other arithmetics which include calculations with fractions, the *Liber algorismi* of John of Seville (ca. 1130-50),[16] which is apparently a translation from a Moslem writer sometime after al-Khwarizmi,[17] and an anonymous algorism, part of which deals with ordinary fractions.[18]

Passing on to the thirteenth century we can list at least five authors who concern themselves with fractions; Leonardo Fibonacci of Pisa in his *Liber abbaci* (1202, revised 1228),[19] Jordanus Nemorarius in the *Demonstratio de minutiis* (a

---

13 Translated by F. Woepcke, *Atti dell' Accad. Pont. de Nuovi Lincei*, vol. 12 (1859), pp. 230-275, 399-438.

14 Edited and translated into German by M. Silberberg, Frankfurt a M., 1895.

15 See note 36.

16 Transcribed by B. Boncompagni, *op. cit.*, in note 7, II, Rome, 1857.

17 This treatise also seems to have been translated into Latin by Gerard of Cremona. See E. Wappler, "Beitrag zur Geschichte der Mathematik" in *Abhandlungen zur Geschichte der Mathematik*, vol. V (1890), p. 159, note.

18 Published by M. Curtze in *Abhandlungen zur Geschichte der Mathematik*, vol. VIII (1898), pp. 1-27, from a manuscript at the Staatsbibliothek in Munich.

19 Published by B. Boncompagni, *Scritti di Leonardo Pisano*, vol. I, Rome, 1857.

shorter summary of which is known as the *Tractatus minuti-arum*),[20] the unknown Master Gernardus in his *Algorismus de minutiis*,[21] Richard of England (d. 1252) in the *Tractatus minutiarum vulgarium*,[22] and finally an unknown author whose treatise begins, "Minutie (sic) est alicuius integri pars. . . ." [23]

In the fourteenth century the steadily growing interest in fractions is maintained. The most important work of this century is probably the *Tractatus de minutiis* of Johannes de Lineriis.[24] His contemporary, Johannes de Muris, must like-

20 G. Eneström, "Das Bruchrechnen des Jordanus Nemorarius" in *Bibliotheca Mathemetica*, 3rd series, vol. 14 (1914), pp. 41-54.

21 G. Eneström, "Der 'Algorismus de Minutiis' des Meisters Gernardus", in *Bibliotheca Mathemetica*, 3rd series, vol. 14 (1914), pp. 99-149. This also includes a comparison with Jordanus.

22 See note 36 for manuscript.

23 Erfurt Stadtbücherei, Amplon. F. 38, ff. 10v-12v. See note 36. The incipit is practically identical with that of the second part of Peurbach's *Elementa Arithmetica*, Wittenburg, 1534, f. 14r, which begins, "Minutia est pars integri..."

24 The controversy over the author of this work is not yet settled. Two editions of it have appeared, both accompanying an algorism by Prosdocimo de' Beldomandi, Padua (1483), and Venice (1540). These editions together with manuscripts cited by A. Favaro, "Intorno alla vita ed alle opere di Pros-docimo de' Beldomandi" in Boncompagni's *Bullettino di Bibliografia e di Storia delle Scienze Matematiche*, vol. 12 (1879), pp. 60-63, have introduced at least two complications into the picture. First, in both editions of the work, and more than one manuscript, the author's name is read, "Liveriis". Second, in the second edition and at least one manuscript of the *De minutiis*, to the author's name is appended "siculus", which would naturally make us question the fact of the author being the astronomer, Johannes de Lineriis of Amiens. M. Steinschneider in the same volume of the *Bullettino* (p. 345 ff.) has pointed out three things in connection with this problem, first, that the variation in spelling arises from "la difficolta (si potrebbe forse dire l'im-possibilità) di distinguere le due lettre *n* e *v* nei vecchi manoscritti latini..." The obvious derivation of "Lineriis" would be "Ligneriis" or "Lignières". In certain manuscripts where Liveriis is read Lineriis is meant. In the second place Steinschneider points out that there existed a John of Sicily, the author of an exposition on the *Canones* of Zarkali. Keeping these two facts in mind we repeat Steinschneider's third point: "Resta a considerare le operare attribute ad un Giov. de Liveriis, il quale in un manoscritto ed in un' edizione e detto *Siculus*. Io non conosco, ne posse indovinare l'origine del nome Liveriis, e sarei inclinato a credere, che non esiste da per tutto, essendo lezione corrotta da Lineriis, e che il Giovanni Siciliano sia stato confuso col

wise be listed among the authors who composed treatises on fractions. Among the Digby manuscripts we find the following: " Explicit tractatulus canonum tabule minutiarum philosophicarum et vulgarium, qui tractatus et tabula composita a magistro Jo. de Muris, Normano, qui eodem anno [1321] complevit plures alios tractatulos (sic.).[25] Johannes also wrote the famous *Quadripartitum numerorum,* which like Marliani's work contains an interesting example of the early use of a type of decimal fractions.[26] There are at least five or six anonymous

Giov. de Lineriis" (p. 318). This belief is given added weight when we notice that the four manuscripts of the *De minutiis* cited by E. Wappler in his "Beitrag zur Geschichte der Mathematik" in *Abhandlungen zur Geschichte der Mathematik,* vol. V (1890), p. 162 note, all have been read Johannes de " Lineriis ". Curtze (M.) in an article "Anonyme Abhandlung über das Quadratum Geometricum" in *Zeitschrift für Mathematik und Physik,* vol. XL (1895), Hist. Abt. p. 165, note, mentions Johannes de Lineriis as the author of the *De minutiis,* whom he distinguishes from John of Sicily.

As to the date of the composition of the work, one manuscript at the Staatsbibliothek at Munich (Cod. Lat. 14684) mentions 1356, which, if the date of composition rather than copying, would appear to be somewhat late for Johannes de Lineriis of Amiens. The explicit of this manuscript runs: "...explicit liber de minuciis anno domini Mccclvj. In vigilia nativitatis domini a magistro Johanne de Linerijs astronomo egregio editus " (fol. 29v.).

Accounts of Johannes de Lineriis can be found in Pierre Duhem, *Système du Monde,* vol. IV, pp. 60-69; Lynn Thorndike, *A History of Magic and Experimental Science,* vol. III, p. 261; M. Cantor, *Vorlesungen über Geschichte der Mathematik,* 2nd edition, vol. II, pp. 126-7.

My references to the *Tractatus de minutiis* in the second section of this chapter will be to the printed edition of 1540. I also examined the *editio princeps* in the Plimpton Library, in which copy a number of pages are missing.

25 Digby 190. The Macray catalogue suggests that the tract on fractions beginning on f. 72r in this codex is also by Jean de Murs. We can tell from the incipit, however, that it is the treatise of Johannes de Lineriis.

26 There is no complete printed edition of the *Quadripartitum.* A. Nagl has transcribed two chapters in an article, " Das Quadripartitum des Joannes de Muris und das praktische Rechnen im vierzehnten Jahrhundert " in *Abhandlungen zur Geschichte der Mathematik,* vol. V (1890), pp. 135-146. L. C. Karpinski has two articles on the *Quadripartitum,* one of which is an analysis of its contents, " The 'Quadripartitum numerorum' of John of Meurs " in *Bibliotheca Mathematica,* 3rd series, vol. 13 (1912-13), pp. 99-114, and the other a note on the history of decimal fractions in *Science,* vol. 45, (1917), pp. 663-665, which includes an excerpt from the *Quadripartitum.* For the dates of his various compositions see Duhem, *op. cit.,* vol. 4, pp. 30-38; Thorndike, *op. cit.,* vol. 3, pp. 294 ff. and the first article of Karpinski.

studies of vulgar fractions additionally credited to the four-teenth century.[27]

The fifteenth century is particularly rich in arithmetics which include sections on fractions. It is marked on the whole by a close connection with the fourteenth century. An excellent example of this relationship is the so-called *Algorismus Ratisponensis,* which has for its second part a discussion of fractions that is an abridgement of the *Tractatus de minutiis* of Lineriis.[28] From the *Algorismus Ratisponensis* are derived at least two works of the fifteenth century, the *Deutsches Rechenbuch* described by Rath[29] and the Bamberger Arithmetic of 1483.[30] We can mention several other treatises of the quattrocento: the popular arithmetic of Georg von Peurbach (1423-61),[31] the German treatise of Widman,[32] the anonymous *Algorithmus novus de integris, de minutiis vulgaribus et de physicalibus* (Cologne, ca. 1500), two Italian arithmetics, that of Pietro Borghi (Borgi)[33] and the more detailed study of Luca Pacioli,[34] and finally the French works of Nicole Chuquet and Jehan Adam.[35] To these can be added at least five anonymous manuscripts.[36]

27 See note 36.

28 For an analysis of the *Algorismus Ratisponensis* consult E. Rath, "Über ein deutsches Rechenbuch aus dem 15. Jahrhundert" in *Bibliotheca Mathematica,* 3rd series, vol. 13 (1912-3), pp. 17-22.

29 *Ibid., loc. cit.*

30 The Bamberger Arithmetic was the second German arithmetic printed. See D. E. Smith, *Rara Arithmetica,* Boston and London, 1908, p. 15.

31 *Elementa Arithmetices,* Wittenberg, 1534. The second part is entitled "De minuciis seu fractionibus vulgaribus" (fol. 14r.).

32 *Behennd und hupsch Recknung uff allen Kaufmanschafften,* Leipzig, 1489. Second edition, Pforzheim, 1500.

33 I have consulted two editions of his *Arithmetica,* both of Venice, 1484 and 1488.

34 *Summa de Arithmetica Geometria Proportioni et Proportionalita,* Venice, 1494.

35 Chuquet, *Le Triparty en la science des nombres,* Paris?, 1484. See L. Thorndike, *Science and Thought in the Fifteenth Century,* New York, 1929, pp. 158-9, for Adam on fractions.

36 I have refrained from mentioning many arithmetics which do not

II. The Algorismus de minutiis

In the first part we have listed chronologically the principal *tractatus* in the history of vulgar fractions. We now proceed to a topical analysis of Marliani's treatise, comparing it with those already mentioned.

The so-called vulgar, common, or composite fraction that Marliani discusses as distinct from, and more general than, the unit fraction (where of course the numerator is unity) or the sexagesimal fraction (where the denominator is sixty) is perhaps derived, as I have suggested in the first section of this chapter, from Hindu sources. In the *Rigveda* there appears the composite fraction three-fourths.[37] Or it might pos-

contain operations with fractions, such as the Salem Codex, the Bernhard Arithmetic, etc.

There are several anonymous or little known works in manuscripts that I would have examined except for the present war. These have been extracted in the order that they appear in Lynn Thorndike's *A Catalogue of Incipits of Medieval Scientific Writings in Latin*, Cambridge, Mass., 1937.

Brit. Mus. Royal Mss. 12 D. VI, c. 1400, ff. 90—(93v) (Th. c. 21).

Cambridge U. Library, 1719 (Ii. I. 27) f. 7, 1424 A. D. (Th. c. 99).

Bodleian Lib. Ashmole Mss. 1522, 14c. ff. 40c-42d. The same ms. in Bodleian Lib. Digby Mss. 28, 14c. ff. 133v(136), (Th. c. 148).

British Mus. Burney Ms. 213, 15c. pp. 349-(369). There is a work of Demetrius Alabaldus (Albaldus) *de minutiis*. (Th. c. 184).

See also N. Bubnov, *Gerberti Opera mathematica*, 1899, p. lxxvii.

Cambridge U. Lib. Ms. 2327 (Mm. III. ii) 15c. f. 18v. (Th. c. 368.)

Bodleian, Digby 17, 15c. ff. 12v-15r (physical fractions) (Th. c. 373.)

Erfurt Stadtbücherei, Amplonian, Q 353, 14cc. ff. 92v-93. (Th. c. 411.)

Bodleian, Tanner Mss. 192, 14c. ff. 18-22v; This is the work of Ricardus Anglicus, also a ms. in British Mus. Royal Mss. 12c. XVII, 14c. ff. 83r-(88r). (Th. c. 411.)

Florence, Biblioteca Nazionale, Palatine, 639, 13c. Apparently a similar incipit to that of Ricardus Anglicus. (Th. c. 411.)

Erfurt Stadtbücherei Amplonian, F. 38, 13c. ff. 10v-12v. (Th. 411.)

Brussels, Bibliothèque Royale, 10126, 14c. (Th. c. 412) (this is probably the tract of John of Lineriis).

Cambridge University 1705 (Ii. I. 13), 14c, ff. 12-14; and also in the same library Gonville and Caius Mss. 141 (191), 14c. ff. 37-53. (Th. c. 589.)

Dresden, C 80, 15c. ff. 157v-166 (fragment) (Th. c. 595).

Vienna, National-Bibliothek, 5498, 15c. ff. 31r-46v. (Th. c. 686.)

37 See Data and Singh, *op. cit.* in note 1, pp. 185-6.

sibly be of Egyptian origin, since the fraction two-thirds is found in the Ahmes papyrus.

For expressing fractions Marliani uses exclusively the early Latin term *minutia* rather than the later *fractio* or *fractus*. Ordinarily after the twelfth century all of these words were apt to be used interchangeably. *Fractio,* which soon displaced the others, is a translation of the Arabic *kasr,*[38] which in turn is a translation of the Sanskrit *bhina,* meaning " broken ".[39] John of Seville, in the *Liber algorismi,*[40] uses *fractiones.* He distinguishes ordinary fractions from sexagesimals by referring to them as " fractions of another denomination ".[41] In the twelfth century Munich manuscript [42] *minutie* is used for sexagesimal fractions, *fractiones* or *minutie diversorum generum* for ordinary fractions. Leonardo Fibonacci of Pisa uses divers terms: *minuta, rupti, fracti,* and *fractiones.*[43] Jordanus employs *minutie* and *fractiones* without discrimination,[44] as does Master Gernardus.[45] Jordanus expresses the difference between sexagesimal and ordinary fractions by the use of different terms. For ordinary fractions the expression *vaga sumptio partium* is employed, while for fractions with fixed denominations (either Roman or sexagesimal), *ordinaria sumptio partium.*[46] Gernardus introduces the distinction that was to become so popular in the fourteenth and fifteenth centuries. He represents sexagesimal fractions as *minutie philosophice* (or *physice*) and ordinary fractions as *minutie vulgares.*[47] Lineriis and many others follow this distinction.

38 Tropfke, *op. cit.* note 8, p. 164.
39 Data and Singh, *op. cit.,* p. 188.
40 *Ed. cit.* in note 16, p. 49.
41 *Ibid.,* p. 56.
42 *Ed. cit.* in note 18, p. 12.
43 *Ed. cit.* in note 19, pp. 23, 27, 71, *passim.*
44 *Ed. cit.,* note 20, passim.
45 *Ed. cit.,* note 21, *passim.*
46 *Ed. cit.,* p. 43.
47 *Ed. cit.,* p. 101. Eneström states that "*philosophice*" is written out in some places while in others the ambiguous "ph'e" is used. There is some

Marliani, on the other hand, in the one chapter in which he has cause to speak of sexagesimals, uses the same term, *minutie,* that he had for common fractions.[48]

Marliani made no formal definition of a fraction. In general definitions were of two kinds, those which followed Euclid in representing a fraction as a part of an integer and those by which it was considered an uncompleted division. The former is by far the most popular and usually appears even with the latter. The conception of the fraction as an uncompleted division was not reintroduced until the twelfth century by the abbacist Odo.[49] We find it appearing again in the thirteenth century in the treatise of Jordanus.[50] Marliani recognizes the connection that exists between fractions and division in his fifteenth chapter: " If you wish to divide an integer by an integer, either the dividend is less than, equal to, or greater than the divisor; if less, you place it above the divisor and you will have a fraction arising from such a division and thus the said divisor will be the denominator and the dividend, the numerator." [51]

To return a moment to the first definition of a fraction, namely as a part of an integer, we note that many times it remains unexpressed in definition but implied and understood in defining numeration and denomination. This is true in the case

confusion in medieval manuscripts between *philosophice* and *physice* since the same abbreviation was almost always used. *Physice* seems to be preferred, perhaps because of the use of sexagesimal fractions in astronomy, but it must be admitted that " philosophical " complements " vulgar " better than does " physical ".

48 *De minutiis*, Paris, BN, Ms., Nouvelles acquisitions latines, no. 761, pp. 13-14.

49 Tropfke, *op. cit.*, I, p. 162.

50 *Ed. cit.*, p. 51. " Si numerus per numerum dividendus sit, scriptis minutiis denominatis a divisore et numeratis a dividendo habebitur dividens." To recognise this of course does not mean abandoning the first definition. In the *Tractatus minutiarum* we find " non aliud esse minutias quam partes. ... Hic quidam secundum relationem ad totum integrum ".

51 *Ms. cit.*, p. 12. " Si volueris integrum per integrum dividere, aut numerus dividendus est minor divisore, aut equalis, aut maior, si minor, illum ponas supra divisorem et habebis minutiam provenientem ex tali divisone et sic dictus divisor erit denominator et dividendus numerator ".

of Leonardo of Pisa, of John of Seville, and of the greater part
of the Moslem and Indian authors. The *Demonstratio de
minutiis* conveys this idea succinctly: " Any multiple part of a
thing is called its fraction or *minutia.*" [52] Gernardus speaks of
a fraction as a quantity numerated and denominated, he having already defined these terms.[53] The anonymous Munich
manuscript of the thirteenth century [54] and the *Elementa
Arithmetica* of Peurbach [55] simply say that a fraction is a part
of an integer. Pacioli says the same thing in Italian: " Rotto
e una e vere piu parti de uno integro ".[56]

While Marliani does not use these definitions, he comes
closest to that of Jordanus when he describes in the first chapter " how any part receives denomination from its *quotenarius,*
or multiple." He concludes that one part of two receives denomination from two, one part of three from three, etc.[57] It is
interesting to note that in this discussion in the first chapter
Marliani treats an improper fraction as perfectly normal. This
is nothing new, for the Indians reckoned with fractions whose
numerators were greater than the denominators.[58] The distinction of " proper " from " improper " fractions is apparently
modern in terminology.[59]

Marliani wrote his fractions as we do today with a numerator above the denominator, separated by a bar. The Indians
wrote fractions in the same way, but without the bar, as early
as 200 A. D.[60] It is in an Arabic work that we first find the
fractional bar used, the arithmetic of al-Hassar.[61] Leonardo

52 *Ed. cit.*, p. 45.
53 *Ed. cit.*, p. 100.
54 Erfurt Stadtbücherei, Amplonian, F. 38, f. 10v.
55 *Ed. cit.* in note 31, f. 14r.
56 *Ed. cit.* in note 34, f. 48r.
57 *Ms. cit.*, p. 1.
58 Tropfke, *op. cit.*, I, p. 165.
59 In chapter VIII Marliani speaks of what we would call an improper
fraction as *minutia reducibilis* (Ms. cit., p. 7).
60 Data and Singh, *op. cit.*, p. 188.
61 H. Suter summarizes the particular passage in *op. cit.* in note 11, p. 24.

Pisano, who introduced it into the West, in his *Liber Abbaci,* undoubtedly borrowed it from Arabic sources, possibly from al-Hassar.[62] On some occasions in Arabic works when the numerator was written above the denominator (without the bar), the integer in cases of mixed numbers was written above the numerator, and hence if there were a fraction without integers, a zero would be written above the numerator.[63]

The acceptance of the fractional bar *(virgula)* in Western Europe after its introduction by Leonardo was by no means universal; it was not used by Jordanus, Gernardus, or in the Bamberger arithmetic of 1483. However, it is found in the arithmetics of Widman, Peurbach, and in the fifteenth century manuscript described by Rath. Pacioli and Borghi likewise employ it, designating it by the Italian term, *vergola.* Marliani, as well as others, accepts the bar without further remarks, it apparently being a common enough device.

Our present day expressions of " numerator " and " denominator " were utilized by Marliani throughout. They date from the early fourteenth century or thereabouts. The earliest reference to them is apparently in the treatise of Lineriis.[64] Before him we find such terms with John of Seville as *denominatio* and *numerus denominationum;*[65] while Leonardo speaks of the numerator as the number *denominans* and the denominator as the number *denominatus.*[66] Jordanus and Gernardus use *numerus numerans* and *numerus denominans.*[67]

---

62 *Ed. cit.* in note 19, p. 24. " Cum super quemlibet numerum quedam virgula protracta fuerit, et super ipsam quilibet alius numerus descriptus fuerit, superior numerus partem vel partes inferioris numeri affirmat; nam inferior denominatus, superior denominans appellatur."

63 See the analysis of the " Satisfying Book " of al-Nasawi by M. F. Woepcke, *op. cit.* in note 9, p. 498.

64 *Ed. cit.*, f. 29r.

65 *Ed. cit.*, pp. 56-57. These same expressions were also used in the anonymous Munich manuscript of the twelfth century. (*Ed. cit.* in note 18, pp. 23-24.)

66 See quotation in note 62, *supra.*

67 Jordanus, *ed. cit.*, p. 45; Gernardus, *ed. cit.*, pp. 100-101.

There is no theoretical difference in the definitions of the numerator and denominator in all of the treatises, only a difference in terminology. Marliani describes the denominator as the number according to which a division is made and the numerator as the number signifying how many of the total parts are used.[68]

With the numerator and denominator clearly defined Marliani then separates common fractions into two types, simple and complex. The simple is the ordinary single fraction *(minutia simplex)* and the complex a fraction of a fraction *(minutia minutie)*.[69] The concept of the complex fraction arises from the second class of the class combinations of the Hindus, the class *Prabhaga* (a/b of c/d of e/f).[70] The Moslem writers knew and employed the complex fraction, calling it a fraction of a fraction *(kesru kesrin)*.[71] Leonardo used it constantly, referring to it in the plural by the " number of fractions under the same bar ". One third of one fourth would be written, according to the Arabic order of right to left, $\frac{10}{43}$.[72] Jordanus and Gernardus do not set aside separate classifications of com-

---

68 *Ms. cit.*, pp. 1-2. " Quandocunque aliquid fuit divisum in aliquot partes equales illud precise componentes, numerus ille secundum quem facta est illius divisio denominator appellatur, et numerus significans quot ex illis partibus accipiuntur, numerator est dicendus ". None of the writers achieved the perfect simplicity of Leonardo Pisano when he declared simply that the superior number affirmed the number of parts of the inferior. See note 62 above.

69 *Ms. cit.*, p. 2. " Scias tamen ut supra quod cum tamen semel sit alicuius divisio, ille partes appellantur simplices minutie illius divisi. Si vero aliquid in partes dividatur equales illud precise componentes, et pars vel partes illarum primarum partium dividantur in alias partes equales, illam vel illas precise componentes, sive bis aut plus fiat divisio, partes ex secunda aut ultiori divisione provenientes, minutie minutiarum appellantur illius primo divisi."

70 Sridhara's *Trisatika*, ed. cit., p. 10; Mahavira's *Ganita-sara-samgraha*, ed. cit., p. 39; Aryabhata II's *Mahasiddhanta*, ed. cit., 146; Bhaskara II's *Lilavati*, ed. cit., p. 6. These are quoted through Datta and Singh, *op. cit.*, p. 190.

71 See al-Hassar, *op. cit., ed. cit.*, p. 24.

72 *Ed. cit.*, p. 59.

plcx fractions, but Lineriis, in a manner quite similar to that of Marliani, explains the reduction of *minutie minutiarum* to *minutie simplices*.[73] Peurbach includes a similar distinction, but strangely enough this comes after and not before the discussion of the general operations with fractions.[74]

In the fourth chapter Marliani explains the reduction of fractions of different denominations to fractions of the same denomination. He constantly uses a phrase current today, the common denominator *(communis denominator)*. Apparently this phrase, like the expressions denominator and numerator, became common in the fourteenth century since it appears regularly in all parts of Lineriis' treatise.[75] Previously we find such phrases as John of Seville's *numerus communis*.[76] In the twelfth century Munich manuscript reduction to a common denominator is explained by " deducere ad inferiorem differentiam ".[77] Jordanus will have us resolve " diversarum denominationum partes in partes similes ".[78] Gernardus speaks of reducing " dissimilium denominationum fractiones ad minucias similium denominationum." [79]

According to Marliani's instructions we must first reduce the complex fractions to simple fractions; then we multiply the denominators of the various fractions together. The result is a common denominator. The denominator of each fraction is divided into the common denominator. The result of that division is then multiplied by the numerator. The product of this multiplication is the numerator while the common denominator becomes the denominator of a fraction which is of

---

73 *Ed. cit.*, f. 30v.

74 *Ed. cit.*, ff. 17v-18r.

75 *Ed. cit.*, f. 29, *passim.*

76 *Ed. cit.*, p. 57. " Numerus communis dicitur, qui ex numeris denominationum duorum laterum in se multiplicatis nascitur."

77 *Ed. cit.*, p. 23.

78 *Ed. cit.*, p. 104, Satz 22.

79 *Ed. cit.*, p. 48.

equal value with the original fraction.[80] This method outlined by Marliani is only slightly changed from the rule cited by all the Hindu authors.[81]

For finding the lowest common denominator almost none of the Western authors offers an explanation. That it was done in the Indian treatises is evident from the discussions of the lowest common multiple and its use in finding the common denominator.[82] The only Western mathematician using the lowest common multiple or the least common denominator appears to be Leonardo Pisano. He speaks of the latter as a *columpna*.[83] The former is expressed in two ways, the *minimum mensuratum numerorum,* and the *minimus commensuraturus omnium numerorum.*[84]

80 *Ms. cit.*, p. 3. " Si primum multiplicabis denominatores omnes inter se, primum scilicet in secundum, provenientem in tertium, si habes tertium, et sic ultra, et proveniens est denominator communis. Hunc communem denominatorem divide per denominatorem prime tue minutie et numerum qui provenit ex divisione multiplica per numeratorem illius tue prime minutie; et numerus proveniens est numerator quem si posueris supra denominatorem communem, habebis unam minutiam que tantum valebit quantum tua prima minutia."

81 Datta and Singh, *op. cit.*, pp. 188-9. Brahmagupta says, for instance : " by the multiplication of the numerator and the denominator of each of the (fractional) quantities by the other denominator, the quantities are reduced to a common denominator." (*Ed. cit.*, p. 172 through Datta and Singh, *op. cit.,* p. 189).

82 Mahavira was the first to speak of the lowest common multiple (*niruddha*). He defines it as follows : " The product of the common factors of the denominators and their resulting quotients is called *niruddha.*" In order to reduce to common denominators he says further : " The (new) numerators and denominators obtained as products of multiplication of (each original) numerator and denominator by the quotient of the *niruddha* divided by the denominator give fractions with the same denominator." Mahavira, *ed. cit.*, p. 33 through Datta and Singh, *op. cit.*, p. 195.

83 *Ed. cit.*, p. 57.

84 *Ed. cit.*, p. 57. He defines the first of the above expressions : " minor numerus qui integraliter dividatur per unum quemque eorum (i. e. numeratorum or denominatorum) " p. 282 he declares : "...2520; qui est minor numerus, in quo reperiuntur omnes rupti prescripti; et vocatur in geometria minimus commensuraturus omnium numerorum..."

Since he did not use the least common denominator in his operations, we would expect Marliani to complete the operations with the common denominator already defined and then, after the completion of the addition or subtraction to reduce the fraction by the elimination of the common factor of the numerator and the denominator. This method of reduction is found in chapters five and six.

The fifth tells us how to find the largest common factor of two numbers. It likewise defines prime numbers. The first thing that must be done is to subtract the smaller number from the larger as many times as possible. There are two possibilities, either there will be or there won't be a remainder. If there is no remainder, then the original smaller number is the common factor. If there is a remainder, it is either unity or some other number. If it is unity, the numbers are prime in comparison with each other, i. e. their only common factor is one. If there is a remainder other than one, this remainder is subtracted from the original smaller number. If nothing remains, the first remainder is the desired factor. If however, there is a remainder, this second remainder is subtracted from the first and the same tests apply. This process can be repeated until either the factor is found or the numbers prove to have no common factor other than one.[85]

In the sixth chapter Marliani uses this common factor of two numbers to reduce fractions. The process, as he names it, is the reduction of a *subtilior minutia* to a *grossior minutia* and can take place naturally when the numerator and the denominator are not *numeri contra se primi*.[86]

---

85 *Ms. cit.*, pp. 5-6. "... subtrahe minorem illorum a maiori quotiens poteris: et facta hac subtractione, aut nihil erit residuum aut aliquid. Si nihil, numerus ille minor est maximus qui utrumque metitur... si aliquid sit residuum vel illud erit unitas vel alius numerus. Si unitas, quoniam si unitas a minori quotiens poterit subtrahatur nihil remanebit, illi duo numeri sola unitate numerantur. Si alter numerus remaneat ille quotiens poterit a minori subtrahatur et post hanc subtractionem, aut nihil remanebit aut aliquid; si nihil, sic numerus qui ante remansit et quem nunc subtraxisti est maximus numeros illos numerans, ... isti duo numeri contra se primi dicantur quod sola unitate numerantur ..."

86 *Ms. cit.*, p. 6.

The reduction of fractions is apparently an elementary operation. Although this operation was not actually included in the Hindu treatises, it was a fundamental process recognized among the Indians. This is shown by a passage in the *Tattvarthadhigama-sutra-bhasya* of Umasvati, a non-mathematical work dated about 150 A. D.: ". . . or as when the expert mathematician, for the purpose of simplifying operations, removes common factors from the numerator and the denominator of a fraction, there is no change in the value of the fraction. . . ."[87]

Leonardo speaks of the common factor of two numbers as the *communis regula*.[88] Gernardus on the other hand uses the same terminology as Marliani, reducing *subtiliores minutie* to *grossiores* by use of a *maximus numerans ambos (numeros)*.[89] Johannes de Lineriis, in his discussion of reduction, uses the expressions *fractiones minores* and *fractiones maiores* as well as the common *minutie subtiliores* and *minutie grossiores*.[90] Georg von Peurbach would reduce a fraction to a *minor numerus*.[91] In the Italian work of di Borghi we find the method of reduction entitled " Como li rotti se schixano ",[92] while that of Pacioli gives us the Latin title *de depressione fractorum sive modo schisandi*.[93] Chuquet, in his *Triparty*, follows precisely the method of Marliani.[94]

Two more types of reduction are given by Marliani before he describes the fundamental operations. The first of these, described in chapter seven, is the reduction of an integer to a fraction of a given denomination. The second, which appears in chapter eight, is the reverse of that process, the reduction of a reducible fraction to an integer.

87 Quoted through Datta and Singh, *op. cit.*, p. 189.
88 *Ed. cit.*, p. 65.
89 *Ed. cit.*, p. 118, Satz 23.
90 *Ed. cit.*, f. 31r.
91 *Ed. cit.*, f. 15r.
92 *Ed. cit.*, f. 41r.
93 *Ed. cit.*, ff. 48v-50r.
94 *Ed. cit.*, p. 604 ff.

To reduce an integer to a fraction we are told to multiply the denominator of a given fraction by the integer and to use the product as the numerator of the desired fraction, the denominator of which is of course given.[95] To reduce fractions where the numerator is greater than the denominator a simple division is recommended.[96]

It was ordinarily not the custom at first to have a special section to point out these forms of reduction. They were included in the descriptions of the fundamental operations with mixed numbers or with integers and fractions. The reduction of integers to fractions is found in the Hindu works in the discussion of the third class (jati) of fractional combinations,, the *Bhaganubandha,* having the form $(a + b/c)$.[97] These types of reduction do not appear as separate operations in Rabbi ben Ezra's *Sefer ha-mispar,* the twelfth century Munich manuscript or Leonardo Pisano's *Liber Abbaci.* It is not until the time of Johannes de Lineriis in the fourteenth century that they are explained as separate topics.[98]

Up to this point we have seen how fractions were defined and described and how the various forms of reduction were discussed. Now it remains to examine the various discussions of the fundamental operations performed with fractions. These operations are the same as those with integers.

Marliani, like many of his predecessors, takes up the addition of fractions first. This is done not with any conception of the psychological value of discussing one operation before another, but simply because in explaining operations on integers addition was the logical place to begin. This order, adapted to integers, was maintained in the case of fractions. However, a

---

95 *Ms. cit.*, p. 7. " ... multiplicabis denominatorem illius minutie per numerum integrorum et proveniens numerus est numerator quem ponam (*sic*) supra denominatorem et habebis minutiam illius denominationis illi equalentem."

96 *Ms. cit.*, p. 8. " ... deinde numeratorem illius proposite minutie per denominatorem suum divide, et numerus ex hac divisione exiens est numerus integrorum in hac minutia contentorum."

97 Datta and Singh, *op. cit.*, p. 193.

98 *Ed. cit.*, f. 31v.

good number of the treatises recognize that multiplication should precede addition. Brahmagupta treats multiplication before addition. Addition is included in his discussion of the first class combination, the so-called *Bhaga,* having the form $(a/b \pm c/d \pm e/f)$.[99] Al-Hassar [100] and Rabbi ben Ezra follow the same order. It was also used in John of Seville's translation, the twelfth century Munich manuscript, Leonardo's treatise, the anonymous German arithmetic described by Rath, the Bamberger arithmetic, and the two Italian works of Borghi and Pacioli.

The method of addition is the same in all treatises, namely the reduction of fractions to a common denominator and then the addition of the numerators.[101] Subtraction likewise is treated universally the same; reduction to a common denominator and the subtraction of the numerators.[102]

In the eleventh and twelfth chapters Marliani notes briefly how to double a fraction by doubling its numerator and halve one by doubling its denominator.[103]

It may be noted at this point that had Marliani composed this work after his treatise on the proportion of motions he scarcely could have refrained from making clear the distinction between doubling and squaring the numerator, the confusion of which he ascribed in that work to the Bradwardine school.

Almost all of the Western treatises include separate sections on doubling and halving. Jordanus and Gernardus recognize two methods of doubling (and halving): either multiply the numerator by two or divide the denominator by two.[104]

99 Multiplication in *ed. cit.*, p. 173; addition in *ibid.*, p. 175.

100 Multiplication, *ed. cit.*, p. 25; addition, *ed. cit.*, p. 29.

101 Marliani says (Ms., cit., p. 8): "...reducas omnes illas ad minutias eiusdem denominationis...omnium numeratores addas insimul..."

102 Marliani, *Ms. cit.*, p. 9. "...reducantur minutie simplices ad eandem denominationem...subtrahe numeratorem minutie subtrahende a numeratore minutie a qua vis facere subtractionem..."

103 *Ms. cit.*, pp. 9-10.

104 Jordanus, *ed. cit.*, p. 49; Gernardus, *ed. cit.*, pp. 106-7.

In chapter thirteen Marliani turns to the operation of multiplication. We have already seen that he makes a distinction between the reduction of complex fractions to simples and the operation of multiplication. This is made evident in this section when Marliani advises us first to reduce complex to simple fractions, for it is only with simples that we are permitted to operate in multiplication.[105]

The method of multiplication that Marliani explains is the common one. Multiply the denominators together and the product becomes the denominator of the result. Multiply the numerators together and place the product above the new denominator as the result's numerator.[106]

This method which originated in India [107] is not the only method of multiplication found in medieval treatises. Al-Hassar, followed by Leonardo Pisano, suggests dividing the product of the numerators by first one denominator and then the other.[108] In Rabbi ben Ezra's *Sefer ha-mispar* we find still a third method of multiplication involving the common denominator. The fractions are reduced to their common denominator. The numerators of the resultants of these reductions are multiplied together, while the common denominator is squared (in the case of three fractions, cubed, etc.).[109] Marliani omits any reference to cross reduction as a complement to multiplication. It existed however from the time of Mahavira.[110]

105 *Ms. cit.*, p. 10. "... reducas illam (minutiam minutiarum) ad minutiam simplicem ... nam solum cum simplicibus licet operari in multiplicando."

106 *Ibid.*, p. 10. "... multiplica denominatorem per denominatorem et productum teneas pro denominatore. Secundo multiplica numeratorem per numeratorem et productum facias numeratorem quem pone supra denominatorem et habebis minutiam provenientem ex tali multiplicatione."

107 Brahmagupta (*ed. cit.*, p. 173) says: "The product of the numerators divided by the product of the denominators is the (result of the) multiplication of two or more fractions." See Datta and Singh, *op. cit.*, p. 176.

108 Al-Hasser, *ed. cit.*, p. 26; Leonardo, *ed. cit.*, p. 59.

109 Ed. cit., p. 33. Ben Ezra also uses as an alternate method that of Brahmagupta.

110 Datta and Singh, *op. cit.*, p. 196.

Following the operation of multiplication in Marliani's treatise is naturally enough that of division. Again Marliani uses the common method, cross multiplication, i. e. the multiplication of the numerator of the first fraction by the denominator of the second and the denominator of the first by the numerator of the second. The first product is the numerator of the quotient and the second is its denominator.[111] This very popular mechanical device of cross multiplication arises from the following rule included in the work of Gernardus: Ordinarily the division of one fraction by another cannot be accomplished with the fractions as they are. Therefore the numerator and the denominator of the dividend are multiplied by the product of the numerator and the denominator of the divisor. When this has been done, a division can be made. The results are as indicated by cross multiplication.[112]

Another common method of division was to reduce to a common denominator and then proceed with the division. This method had its origin with the Hindus.[113] It was likewise used by the Arabic authors,[114] and by Rabbi ben-Ezra, who adds the comment that it is not often necessary to divide fractions.[115] Several Western mathematicians use it also: John of Seville,[116]

---

111 *Ms. cit.*, p. 11. "... multiplica denominatorem minutie dividende per numeratorem minutie dividentis et proveniens teneatur et serva pro denominatore. Secundo multiplica numeratorem minutie dividende per denominatorem minutie dividentis et proveniens tene pro numeratore quem ponas supra denominatorem quem servasti et habebis minutiam exeuntem ex tali divisione."

112 Gernardus, *op. cit.*, pp. 113 and 145. This can be shown symbolically as follows:

$$\frac{\dfrac{a}{b}}{\dfrac{c}{d}} = \frac{\dfrac{acd}{bcd}}{\dfrac{c}{d}} = \frac{ad}{bc}.$$

113 Datta and Singh, *op. cit.*, pp. 198-199.

114 See H. Suter's analysis of al-Nasawi's Arithmetic in *Bibliotheca Mathematica*, 3rd series, vol. 7 (1906-7), p. 114.

115 *Ed. cit.*, p. 39.

116 *Ed. cit.*, pp. 70-71.

the author of the anonymous Munich manuscript,[117] and Leonardo Pisano.[118]

A third method of division, the one which we use today, is to invert the divisor and then to multiply. Tropfke makes the erroneous statement that this appeared for the first time in Stifel's *Arithmetica integra* (1544).[119] Actually it is one of the oldest methods, having been employed by the Indians,[120] and by the Arabs.[121]

In the operations of multiplication and division some medieval mathematicians without a clear perception of the nature of fractions introduced certain logical difficulties. Why is the result of the multiplication of fractions less than the multipliers? And in division, why is the quotient greater than the dividend? (These questions were of course asked of proper fractions). Marliani does not pose these questions, but a notation on the margin of page twelve of the Paris manuscript asks us to consider them.[122] Jordanus explains a similar difficulty with respect to sexagesimal fractions.[123] The kindred problem of the root being greater than the number from which it is

117 *Ed. cit.*, p. 24.

118 *Ed. cit.*, p. 71 ff.

119 Tropfke, *op. cit.*, I, p. 170.

120 Brahmagupta tells us, *ed. cit.*, p. 173: "The denominator and the numerator of the divisor have been interchanged; the denominator of the dividend is multiplied by the (new) denominator and its numerator by the new numerator. Thus division of proper fractions is performed." See definitions also by Sridhara, and Mahavira. Datta and Singh, *op. cit.*, p. 198.

121 See Suter's analysis, *loc. cit.* in note 114.

122 *Ms. cit.*, p. 12. "Considera quare in minutiis numerus multiplicans et numerus multiplicatus sit (*sic*) maior numero producto, et similiter numerus proveniens ex divisione sit maior diviso, cuius oppositum est in integris, quoniam videtur contra illud principium omne totum est maius sua parte."

123 Jordanus, *ed. cit.*, p. 44. "Ex vi relationis, quoniam gradus in gradum continuo facit gradum, minutum in se faciet secundum, sed in respectu gradus producti. Per se enim quelibet minutia, hoc ex natura supposite quantitatis considerata, integrum est, et integrum simile producit, hoc est minutum facit. Cum vero ex partis multiplicatione producitur secundum, non minus quidem quantitate sed minus secundum proportionem quantitatis."

extracted is found in a fragment, *De diversitate fraccionum,* contained in a Dresden manuscript.[124]

We have arrived finally at the last chapter of Marliani's treatise. This is entitled "how the root or approximate root of any number is found." This method of root extraction involves an early form of decimal fractions. The method is briefly this: If the number of which the root is to be extracted is not a mixed number, but a whole number by itself, we add to it twice as many zeros as the number of places to which it will be carried. Then the root is extracted by the ordinary rule of algorism. From the answer we then remove as many figures as one half the number of zeros originally added. The remaining figures represent the integral part of the desired root, while of course the figures removed represent a fractional part having a denomination of the power of ten equal to half the number of zeros added. The figures removed (the fractional part) are converted into sexagesimal fractions (from decimal fractions) by the simple procedure of multiplying them by sixty, removing the same number of figures as before, and hence having as a result the number of minutes. These figures removed in turn may be multiplied by sixty to find the number of seconds, etc.[125] The example that Marliani

124 Ms. C 80 analyzed by E. Wappler in his article "Beitrag zur Geschichte der Mathematik," *Abhandlungen zur Geschichte der Mathematik,* vol. 5 (1890), p. 161, note 2. It is in the explicit of this fragment that we note the above problem: "...non est autem mirandum quod maior est numerus radicis quam quantitatis cuius est radix, hoc enim accidit in fraccionibus sed non in integris. Radix enim per multiplicacionem queritur, sed in multiplicatione integra crescunt, fractiones vero descrescunt quantitate licet eorum numerus augeatur sicut habitum fuit supra."

125 *Ms. cit.,* p. 13. "Cum alicuius numeri radicem quadratam veram vel saltem illi satis propinquam invenire volueris, si non fuerit cum numero aliqua minutia, illi propones multas cifras et quanto plures tanto precisius secundum tantum numerum partem et non aliter. Deinde illius ita composite accipias radicem quadratam secundum regulas algorismi, et a radice quadrata sic inventa reconde ex primis figuris tot figuras quot fuerit medietas cifrarum additarum, et remanentes figure sunt radix propositi numeri in integris. Si aliquas figuras significatas amovisiti post multiplica eas figuras quas primo amovisti per 60, et a provenienti remove etiam tot figuras sicut prius et ex primis et remanentes figure significant quot minutie ultra integra contineantur

gives is the extraction of the square root of eight. It can be represented thus:

$$\sqrt{8} = \frac{1}{100}\sqrt{80000} = 2.82 = 2°\ 49'\ 12''.$$

This method of root extraction grew out of a similar usage of sixty for a base instead of ten. It appears to have been used first by the Arabic authors. Al-Nasawi gives two examples that show clearly the origin of this early form of decimal fractions: [126]

1. $\sqrt{26°\ 17'} = \frac{1}{60}\sqrt{94,620''} = \frac{1}{60}\,307' = 5°\ 7'.$

2. $\sqrt{17°} = \frac{1}{100}\sqrt{170,000°} = \frac{1}{100}\,412° = 4°\ 7'\ 12''.$

The same method is used by John of Seville, Jordanus, John of Murs, and John of Gmunden.[127]

After his explanation of root extractions Marliani's *De minutiis* breaks off without any formal conclusion. It is quite probable that two chapters are missing, or at least that they were in the original plan of the work, chapters on the extraction of the square and cubic roots of fractions. Almost all treatises include these two operations.

In summary we can say that Marliani's treatise contains no original material, but that it is not a slavish copy of any hitherto known work. Resembling most closely the discussions of Johannes de Lineriis and Georg von Peurbach, it is simply written with the rules tersely stated without the unnecessary repetition found in so many of the treatises. Marliani does not inject his personality into this work as he does in his physical studies. In addition, this short piece is free from the argumentation and scholastic form of the others.

in radice. Deinde figuras quas amovisti multiplica etiam per 60, et a provenienti remove ex primis figuris tot sicut prius et remanentes figure significant quot secunde in radice ultra integrat et minutias."

126 H. Suter, "Über das Rechenbuch des Ali ben Ahmed el-Nasawi" in *Bibliotheca Mathematica*, vol. 7 (1906), p. 118.

127 George Sarton includes these examples in an article, "First explanation of the decimal system. Together with a history of the decimal idea," in *Isis*, vol. 23 (1935), p. 168 ff.

CONCLUSION

MARLIANI's compositions happily fall into three natural divisions, the first two of which are physical and the last mathematical. The first of these divisions is concerned generally with heat actions and includes three problems in particular:

(1) Reaction, i. e. how a heating agent is cooled by the cold patient on which it acts. In discussing this question Marliani presents a broad picture of late medieval concepts of hot and cold as argued in the English, French, and Italian universities. His constant use of numerical degrees of intensity of heat and cold shows a familiarity with the temperature concept, from which he distinguishes the idea of quantity of heat and cold. From the doctrine that every body has a goal or perfection of heat or cold reception beyond which it can not be heated or cooled (the so-called *summus gradus caliditatis* and *summus gradus frigiditatis*), he deduces a temperature scale of a fixed number of eight degrees to measure imperfect reception or attainment of the goal. When a body is said to be imperfectly hot to a certain degree, it is at the same time said to be imperfectly cold to a fixed degree dependent on its intensity of heat. In this way a degree of intensity of heat has a complementary or dependent coextensive degree of intensity of cold. This is a doctrine that would lead ultimately to the abandonment of the dual concept of hot and cold.

(2) The cooling of heated water. In this study Marliani emphasizes the same ideas to describe heat action as in the preceding discussion. He believes strongly that heated water must be cooled by contact with the air that surrounds it, which is colder, and not by any intrinsic property of the water, as was commonly believed by medieval schoolmen. His use of a controlled experiment to prove that boiling water actually freezes more rapidly than non-heated water is of some interest, although he is unaware that the cause of this phenomenon lies chiefly in the decreased mass of water resulting from the rapid evaporation of the boiling water.

(3) Body heat and temperature. Following his Italian predecessors Marliani here distinguishes between body temperature and quantity or production of natural body heat. He partially accepts John of Sermoneta's theory of the constancy of body temperature regardless of the season of year. He believes at the same time that the production of natural heat is higher in the winter than in the summer. His discussion of antiperistasis, the supposed intensification of a quality as a result of contact with its contrary quality, is not very satisfactory.

The second division of Marliani's works contains two problems of mechanics. (1) The first, a study in kinematics, is concerned with finding a uniform velocity with which a body can traverse the same amount of distance in a given time as a body moving with a uniformly accelerated velocity. He proves anew the conclusion of the Calculator (Swineshead) and Dumbleton, which had held that the desired uniform velocity is equal to the mean between the initial and final velocities of the uniformly accelerated movement. His proof is inferior to the geometric proof of Nicolas Oresme, but is clearer, if less general, than those of his English forerunners. (2) The second problem is a study of the direct proportionality law of motion, which described velocity as following the ratio of the motive power to the resistance. Marliani rejects the law because of its mathematical inconsistency, i. e. when the motive force is equal to the resistance, there is no motion, as Aristotle and the schoolmen would agree, but, according to the law, a definite velocity will result when the force equals the resistance (or in other words when the ratio is one). Marliani, however, does not interpret rightly Bradwardine's correction of the law, misunderstanding what the latter meant by double the proportion, composition of proportions, and proportion of proportions.

Both these physical divisions of heat action and local movement are united in Marliani's commentary on Richard Swineshead's *Liber calculationum.*

The third and final division includes Marliani's mathematical studies, only one of which, the *Algorismus de minutiis,* is extant. This work is a simple and lucid treatment of fractions. But like all the medieval discussions it is no improvement in theory on the original Hindu treatises. It does, however, have some interest for us because it employs terms and definitions commonly used today.

In conclusion, it should be noted that Marliani, as a physicist, was superior to most of his contemporaries, but that he was less important, from the standpoint of modern science, than his English and French precursors, in that he seems rarely to get beyond the doctrines of those precursors. This conclusion seems to bear out the pessimistic judgment by both Duhem and Thorndike of the fifteenth century science in comparison with that of the fourteenth.

APPENDIX I
SWINESHEAD

THERE has been considerable controversy over the identity and the works of the mysterious English philosopher or philosophers of the fourteenth century, bearing the name Swineshead. In the center of this controversy is the question whether the variance of Christian names assigned to Swineshead is an indication that there was more than one schoolman named Swineshead, or merely that scribes in copying manuscripts altered the one original Christian name accompanying the cognomen Swineshead or its variations (Swyneshede, Suiseth, Suisset, Swicet, etc.). The contemporary evidence is too slight to solve this question once and for all. However we can collect and examine the essential facts known to us at this time, and indicate some possible conclusions. In doing so we must dismiss, or at least set aside for lack of confirmation, some of the material appearing in accounts of Swineshead composed in the sixteenth, seventeenth, and eighteenth centuries, and rely as far as possible on contemporary evidence or on material derived from the manuscripts and various editions of the tracts attributed to Swineshead. The following considerations stand out as the most important:

(1) A treatise on motion is attributed to Swineshead, of which there is one complete extant manuscript, Erfurt, Stadtbücherei, Amplonian F. 135, ff. 25-47. This manuscript, which is dated 1333 A. D., designates the title of the work as *De motibus naturalibus et annexis*. In addition to this manuscript there is a detailed discussion of the treatise by a Parisian student of the fourteenth century which is preserved in the Bibliothèque Nationale in Paris, MS Fonds latin, 16621, f. 35ff. He names the work from its incipit *De primo motore*. In both manuscripts the Christian name Roger is used with the cognomen Swineshead, although apparently through an error " William " appears once in the Erfurt manuscript. Bale

knew of this treatise as the *Descriptiones motuum*, citing its incipit from a copy in the library of Thomas Godsalua.[1] Whether this copy cited by Bale had the praenomen Roger we can not tell from his description, but he seemed to believe that Swineshead was named Roger, for so he designates him. There is, however, no reference to a Roger Swineshead in the fourteenth century documents of Oxford.

(2) Another work attributed to Swineshead is a *Lectura* or *Questiones* on the *Sentences* of Peter Lombard. The attribution of this treatise to Swineshead, so far as we know, appears in only one manuscript, MS. Oriel College 15.[2] I have not been able to examine the *Questiones*, but Michalski, after reading the copy noted above, identifies the *Questiones* with the *Lectura* on the *Sentences* written by a Roger Rosetti.[3] He cites Rosetti manuscripts of the Bibliothèque Publique de Bruges (MS. Bruges 192) and the Vatican Library (Cod. Lat. 1108). It is difficult to tell whether Michalski read either of these latter manuscripts in identifying them with that of Swineshead. At any rate their incipits are the same.[4] Now Michalski makes the assumption that because the Rosetti manuscripts are identical with the Swineshead manuscript, that the name Rosetti must be a variation of Swineshead. This is not so. Roger Rosetti is actually a variation of Roger Royseth, who was an English Franciscan.[5] The Bruges and Vatican manuscripts are copies

1 John Bale, *Index Britanniae Scriptorum*, Oxford, 1902, p. 403.

2 Coxe, H. O., *Catalogus codicum manuscriptorum in collegiis aulisque oxoniensibus*, Oxford, 1852, I, MSS. of Oriel, p. 6.

3 K. Michalski, "Le Criticisme et le Scepticisme dans la Philosophie du XIVᵉ siècle," in *Bulletin International de l'Academie Polonaise des Sciences et des Lettres*, Classe de philologie. Classe d'histoire et de philosophie, L'Année 1925, Part I, pp. 47, 79.

4 "Utrum aliquis in casu ex precepto possit obligari ad aliquid..." *Cf.* Coxe, *op. cit., loc. cit.* and *Cod. Vaticani Latini* (Edited by Pelzer), Vatican, 1931, v. II, Cod. 1108. The latter catalogue identifies the Vatican and Bruges manuscripts.

5 Giovanni Sbaraglia, *Supplementum et castigatio ad scriptores trium ordinum S. Francisci*, Rome, 1806, p. 647, gives an account of Royseth in

of Royseth's treatment of the *Sentences*.[6] If then the Oriel
manuscript is identical with those of Bruges and the Vatican,
as it seems to be, then we must conclude that the attribution
to Swineshead of *Questiones* on the *Sentences* is false.

(3) A third work assigned to Swineshead is a *Liber calculationum*. There are several extant manuscripts and editions. The
manuscripts are: Venice, BN San Marco VI, 226; CU Gonville and Caius 499 (incomplete); Worcester Cathedral F35
(incomplete); Paris, BN Fonds latin 6558; and Rome, Vittorio Emanuele 250.[7] The editions are: Padua, ca. 1477;
Pavia, 1498 (apparently the edition of Pavia, 1488 is an
error); and Venice, 1520. All of the manuscripts and editions
use the name Swineshead in one form or another. It is true
that the form in the Paris manuscript, Ghlymi Eshedi, is not
particularly close to the original. It was because of this variance
that Duhem chose to believe that the author of the *Liber calculationum* was not named Swineshead.[8] But it is quite obvious,
as Michalski and Thorndike have suggested, that Ghlymi
Eshedi is a copyist's error for Swyneshede.[9]

The only manuscript to include a Christian name is that of
Paris, which calls Swineshead (i. e. Ghlymi Eshedi) Richard.[10]
He is likewise named Richard in the editions of 1498 and
1520. There is no Christian name in the *editio princeps*. We
can probably dismiss the name " Raymond " which appears in

which he cites a manuscript of the *Questiones* on the *Sentences* from an
inventory of 1381 of the Convent of Assisi.

6 The incipit cited by Sbaraglia of the work from the 1381 inventory is
identical with the Bruges, Vatican, and Oriel manuscripts. Pelzer's Vatican
catalogue identifies the Bruges and the Assisi manuscripts with a Cusan
manuscript.

7 These manuscripts have all been cited by L. Thorndike in his *A History
of Magic and Experimental Science*, III, pp. 372-3. I have checked them in
the various catalogues with the exception of the Vittorio Emanuele manuscript. I have read the Paris manuscript as well as the *editio princeps*.

8 *Études*, III, pp. 418-20.

9 Thorndike, *op. cit.*, III, pp. 373-4; Michalski, *op. cit.*, p. 61.

10 Vittorio Emanuele 250 might be an exception. I have been unable to
check either title or incipit of this copy.

the colophon of the edition of 1520 as an error, since in the title of that edition Richard has already been used.

A Richard Swineshead appears in the Merton College records of 1339, and was known to have taken part in the election of Wylliott as chancellor of Oxford in 1349.[11]

(4) The fourth important treatise believed to have been composed by a Swineshead is a logical tract of two parts, *De obligationibus et insolubilibus*. The Vatican library has several manuscript copies: Cod. Vat. Lat. 950, f. 117ff.; 2185, f. 78r-v; 3065, ff. 122v-125r; 2154, ff. 1r-6r, 6r-12v; 2130, ff. 152r-154v. The printed catalogue contains a complete description of no. 950 only. In this description the other manuscripts are mentioned.[12] Manuscript no. 950 has neither praenomen nor cognomen. From the citation of the others it is impossible to tell how or whether the author's name is mentioned in them. A Venetian manuscript entitled *Tractatus obligationum* (San Marco, Z. L. 301, f. 41-44) carries the name of Swineshead (Suiseyt) without any Christian name.[13]

There is a Bodeian manuscript of the logical tracts of Swineshead with no author's name (Ms. 2593).[14] At the time Bale composed his Index there was a copy of *De insolubilibus* with a *De divisionibus* at Magdalen College, Oxford.[15] Such a manuscript is not mentioned in the later catalogues of Bernard and Coxe. Bale likewise mentions a *Sophismata logicalia* in the library of Thomas Godsalua. However, since none of these manuscripts mentioned by Bale are extant as far as we know, they can offer us no evidence as to their author's Christian name.

11 G. C. Brodrick, *Memorials of Merton College*, Oxford, 1885, p. 213. A. Wood, *Historia et Antiquitates Universitatis Oxoniensis*, v, I, Oxford, 1674, p. 171.

12 *Codices Vaticani Latini* (Edited by Pelzer), vol. II, Cod. 950, p. 395.

13 Valentinelli, *Bibliotheca Manuscripta ad S. Marci Venetiarum*, Class XI, Cod. 12.

14 F. Madan, *A Summary Catalogue of Western Manuscripts in the Bodleian Library at Oxford*, vol. II, part 1, Oxford, 1922, p. 122.

15 Bale, *op. cit.*, p. 404.

Finally among these logical manuscripts of Swineshead we can cite a Parisian manuscript of the *Obligationes* (BN, 14715, ff. 86-90). According to this manuscript the treatise was composed by the Reverend Jo. (Joannes) Swiinsed, a Doctor of Theology. The whole codex in which this appears is dated 1378.[16]

There is a John Swineshead who appears as a Fellow of Merton College in 1346,[17] who might conceivably be the Swineshead referred to in the Paris manuscript of the *Obligationes*.

I shall omit any discussion of the additional fragments on motion in CU Gonville and Caius 499, ff. 204-215 and Bodleian Digby 154, ff. 42-46r, which may be parts of the *Liber calculationum* or of *De motibus naturalibus*.[18] They give no information as to the Christian name of their authors.

Also of no value in settling the identification problem is the knowledge furnished us by Bale that there were two manuscripts in his day of a Swineshead commentary on Aristotle's *Ethics,* since we are now unable to locate either manuscript.[19]

Now that we have listed the principal facts known about Swineshead, what conclusions are suggested by them? There seem to be two general conclusions possible. Neither is very conclusive, but the second is closest to the evidence and is more likely.

(1) There was only one Swineshead, a philosopher and logician, who was at Oxford in the first half of the fourteenth century. Perhaps he is the Richard of Merton College. His last name suggests that he might have been a Cistercian monk from the monastery of Swineshead. The omission of the Christian name in so many manuscripts raises the possibility that

16 Duhem, *Études*, III, p. 413.

17 Brodrick, *op. cit.*, p. 212.

18 See L. Thorndike, *op. cit.*, III, p. 376. Consult also the James catalogue of Gonville and Caius. The Digby manuscript (cited in Macray's catalogue) bears neither praenomen nor cognomen, but has the same incipit as one of the fragments of the Gonville and Caius manuscript.

19 Bale, *op. cit.*, pp. 403-404.

he was known well enough in the fourteenth century for his last name to identify him.

(2) There was more than one Swineshead, possibly three, Richard, Roger, and John. In support of this we have the variations in Christian name in the various manuscripts. Of the three, Roger is the only name to appear in copies of *De motibus naturalibus,* and Richard, in copies of the *Liber calculationum.* We can not be as sure that John is the only name associated with the logical tracts, for we have been unable to read the incipits and explicits of some of the various Vatican manuscripts. Richard and John are mentioned in contemporary records. The name Swineshead would be common enough, since there was a Cistercian monastery of that name.

INDEX

Actions, heat, 34, 36 ff., 41 ff., 58, 67, 69, 73-74, 77-78, 129, 168-69
Adam, Jehan, 150
Adiuta, Philip, 20, 27-28, 32, 138-39
Agent and patient, 36-39, 41 ff., and see Actions
Ahmes papyrus, 152
Air, 42, 45, 52, 59, 61-63, 72-73, 76, 83, 93-94, 96-98, 168, middle region of, 52, 93, 96, 100
Alabaldus, Demetrius, 147, 151
Albert of Saxony, 17, 40, 42-46, 50, 54, 59-61, 133-36, 138-39, 144
Albertus Magnus, 19, 130
Alexander of Aphrodisias, 19, 81, 91, 94
Algebra, 145
Algorism, 147-48, 166
Algorismus Ratisponensis, 150
Algorithmus novus de integris, etc., 150
Al-Hassar, 146, 154-56, 162-63
Al-Kalcadi, 147
Al-Karkhi, 146
Al-Khwarizmi, 29, 146-47
Alkindi, 35, 133
Al-Nasawi,, 146, 155, 164-65, 167
Amodeo, F., 121
" Ancients," 72
Animal, 64
Antiperistasis, 31, 52, 79, 81, 86, 91, 92-100, 169
Antiquity, 34, 101
Arabia and Arabic, 94, 129, 145-47, 152, 154-56, 164-65, 167
Arcolani, Giovanni, 20, 25, 28, 60, 65, 66-77, 97, 138
Argellati, Filippo, 11 ff., 25, 31-32
Aristotle and Aristotelian, 21, 36, 43, 52, 58-59, 65, 125, 138, 142, *Caelo*, 126-27, *Ethics*, 175, *Generation and Corruption*, 39, 46, *Mechanica*, 125, *Meteorologica*, 72, 93-94, *Physics*, 23-24, 49, 53, 61-62, 63, 107, 125, 127-30, 139, *Problemata*, 80, 91, 94
Arithmetic and arithmetical, 102, 136, 141, 145-46, 150, 154
Aryabhata II, 146, 156
Astrology, 15, 17-18
Astronomy, 21, 145
Auctor de sex inconvenientibus, 40, 42, 50, 104, 135
Averroes, 129-30, 142
Avicenna, 31-32, 59-60, 62-65, 68, 72-73, 80, 89, 91, 94

Bale, John, 171-75
Bamberger Arithmetic, 150, 155, 162
Barry, Frederick, 7
Baths, thermal, 64, 98
Bauer, L., 34
Beldomandi, Prosdocimo de', 148
Bernard, E., 174
Best, C. H., 83
Bhaskara II, 145-46, 156
Bigongiari, Dino, 7
Björnbo, A., 29
Blasius de Parma, 120-21
Blood, 82, 85, 87
Body, human, 79-92
Bologna, University of, 15, 27, 65, 80
Boncompagni, B., 146-47
Borchert, E., 121
Borghi, Pietro di, 150, 155, 160, 162
Bosso, Donato, 21
Boutaric, A., 34
Boutroux, P., 126
Bowels, 81
Boyer, C., 7
Boyle, Robert, 93
Bradwardine, Thomas, 7, 17, 105, 130-40, 144, 162, 169
Brahmagupta, 145, 158, 163, 165
Brain, 85
Brambilla, Joh. Alex., 14
Brodrick, G. C., 174-75
Bruges, Bibliothèque Publique of, 172-3
Bubnov, N., 151
Buridan, Jean, 7, 38, 50, 80, 98, 125, 135, 139
Burley, see Walter of

Cajori, F., 34
Calculator, see Swineshead, Richard
Calidity, see Heat and Cold
Caloric, 34, 44
Cambridge, University library, 151, Gonville and Caius College library, 173, 175
Camendzind, C., 126
Campanus of Novara, 133, 139
Canterbury, archbishop of, 131
Cantor, M., 149
Capitani of Milan, 15-16
Cardomomum, 35
Casali, Giovanni, 40-42, 44
Castello di Porta, 15
Center of gravity, 98
Charles V, German emperor, 11
Chemical, 83

Choleric people, 88
Chuquet, Nicole, 150, 160
Clienton, Richard, 140
Clouds, 93-94
Codice diplomatico of the University of Pavia, 14 ff., 80
Cold, see Heat and,
Colebrooke, H. T., 145-46
College of Physicians, Milan, 11, 14
Collegio Marliani, Pavia, 12
Commentator, see Averroes
Complexio, 77, 85, 89-90, 92, 97
Composition of proportions, see Proportions
Compounds, medicinal, 35
Conciliator, see Peter of Abano
Conduction of heat, 45
Confaloneriis, Damianus de, 30, 32
Congelation, 52, 76-77, 96
Constantine the African, 33-36
Constellations, 61
Contraries and contrariety, of qualities, 36-38, 41 ff., 51-53, 59, 61-62, 64, 71, 73, 77, 81, 92, 94-100, 169
Convulsions and spasms, 28
Coordinates, geometric, 102, 120
Coronel, Louis, 104
Corte, B., 12, 14, 20-21, 29
Costa ben Luca, 82
Coxe, H. O., 172, 174
Curtze, M., 147-49

Dallari, U., 27, 80
Datta, B., and A. N. Singh, 145, 151-52, 156, 158, 160-65
De proportionalitate (proportione) motuum et magnitudinum, 101, 130
Decimal fractions, 149, 166-67
Degrees, temperature, 34 ff., 47, 55-58, 62, 63, 66-69, 71-74, 77, 85, 87, 168, of motion (velocity), 103-124
Demons, 100
Denomination of a proportion, 133, 140-41
Denominator, 151 ff., least common, 158
Density, 34, 48, 57, 63, 70-71, 76-77, 100, 127-28
Descartes, 120
Deshayes, M., 126
Digestion, 28
Dijksterhuis, E. J., 125-26
Diodoros, 93
Distance, see Space
Division, of integers, 153, of fractions, see Fractions
Dominium, 130, 135
Dresden, ms. on fractions at, 151,

166
Duhem, P., *Études sur Léonard de Vinci*, 7, 23-25, 30, 38, 40-41, 101-02, 104-05, 120-21, 125-26, 130-31, 135, 170, 173, *Système du monde*, 40-41, 149
Dumbleton, John, 7, 102, 117-19, 124, 133, 135, 139-41, 169
Dvivedi, P. S., 145-46
Dynamics, 125-44, Aristotle's, 125-29, medieval, 129-44, of Marliani, 141-44

Egidius Romanus, 42, 60, 61, 130
Elements, 43, 66, 68, 73
Empedocles, 46
Eneström, G., 101, 148, 152
Environment, 84, 87, 90
Erfurt, Stadtbücherei at, 148, 151, 154, 171
Ethiopia, 94
Euclid, 133, 139, 153
Euglyphus, Johannes, 19
Evacuation of crude matter, 28
Evans, A. P., 7
Evaporation, 72, 168, and see Vapors
Experience, 43-44, 62
Experiment, 43, 47-49, 54-55, 64, 72, 76, 89, 96, 99, 140

Fall of bodies, 127, law of free, 101
Favaro, A., 148
Ferrara, University of, 65-66
Ferrari, H., 18, 20
Ferrari da Grado, Gian Matteo, 18, 20, 32
Fever, 28, 31, 90
Fibonacci, Leonardo, 147, 152, 154-65
Filelfo, Francesco, 19
Fingers, 91
Fire, 51, 63, 73, 99
Florence, Biblioteca Nazionale at, 29, 151, Biblioteca Riccardiana at, 40, University of, 80
Food, 81
Force, in Newtonian sense, 125-26, and see Powers
Form, 39, 59-61, 65, 71, 75, 89, 94-95, 108
Formentini, M., 18
Fractions, 145-67, 170, addition and subtraction of, 161-62, decimal, 166-67, defined, 153-54, division of, 164-65, multiplication of, 163, reduction of, 159-61, sexagesimal, 146, 149, 151-53, 165-67, terms used for, 152-53, use of bar in, 154, use of common denominator in, 157-58

Freezing and frozen, 59, 71-72, 76-77, 93-94, 95, 100, 168
Frigidity, see Heat and Cold
Furnace, 99

Gaetan of Tiene, 20, 23-25, 32, 41-42, 47, 51-56, 95-96, 99, 121
Galen, 34, 68, 80, 90
Galileo, 22, 101, 120
Gandz, S., 145
Gaza, Theodore, 81
Generation, 85, 89
Gentile da Foligno, 36, 80, 89
Geometry and geometric, 77, 132, 136, 141, Oresme's coordinate, 102, 120-21, 169
George of Trebizond, 81
Gerard of Brussels, 101
Gerard of Cremona, 29, 147
Gerbert, *Opera mathematica*, 151
Gernardus, Master, 148, 152, 154-57, 160, 162, 164
Geyer, B., 40
Gherardi, A., 80
Ghilini, Girolamo, 31
Giacomo delle Torre, See Jacobus de Forlivio
Godsalua, Thomas, library of, 172, 174
Gordon, Bernard, 36
Greeks and Greek, 81, 92, 129
Griffo, Ambrogio, 19-20, 81, 99
Grosseteste, Robert, 34

Hail and hailstones, 52, 93-95, 100
Hammerle, K., 126
Hand, 44, 49, 55
Head, 44
Heat (hot) and cold, 34 ff., 41 ff., and Caps. II-IV, body, 79-92, 94, 96, conduction of, 45, intensity of, 35, 37-38, 42, 47, 51, 56-58, 63, 65, 70, 74, 78-79, 82, 87, 99, 168-69, quantity or extension of, 35 ff., 41 ff., 47, 55-56, 58, 60, 63, 65, 67, 70, 74, 78-79, 82-83, 87, 168-69
Heavens, 45
Heidelberg, University of, 40
Hentisbery, William, see Heytesbury
Heytesbury, William, 19-20, 49-51, 102, 111-12, 121, 140-41
Hindu mathematics, see Indian mathematics
Hindu numbers, 146
Hippocrates, 21, 80-81, 83, 88
Hochheim, Ad., 146
Hugh of Siena, 80, 84, 88
Hugo Senensis, See Hugh of Siena
Human beings, 49

Ibn al-Banna, 146
Ibn Ezra, Abraham, 147, 161-64
Ice, 59, 61, 72, 93
Impetus theory, 50, 125-26
Indian mathematics, 145-46, 151, 154, 156, 158, 160-65, 170
Inertia, 125-26
Infinitesimals, use in kinematics, 107, 110, 114, 122, 124
Infinity, 107, 122
Instant, Peter of Mantua and Marliana on the, 27, 33, 106, concept of velocity at an, 107, 110, 114
Inventario dei manoscritti (of the University of Pavia), 25-26
Iron, 43, 46

Jacobus de Forlivio, 38, 40-43, 47, 50, 54, 60, 64, 77, 79, 81-85, 88, 94-95, 99, 104-05
James of Forli, see Jacobus de Forlivio
James, M. R., 175
Jansen, B., 126
Johannes de Arculis, see Arcolani, Giovanni
Johannes de Lineriis, 148-50, 155, 157, 160-61, 167
Johannes de Sermoneto, see Sermoneta, Johannes
John of Gmunden, 167
John of Lignières, see Johannes de Lineriis
John of Murs, 135, 148-49, 167
John of Seville, 147, 152, 154-57, 162, 164, 167
John of Sicily, 148-49
Jordanus of Saxony, see Nemorarius, Jordanus
Jordanus de Turre, 36

Karpinski, L. C., 149
Kaye, G. R., 145
Kibre, P., 36
Kinematics, fundamental theorem of, 101-24, 130-31, 169

Lantzsch, C., 35
Latitude of motion, 103-24, of qualities, 37, 73-74, 103
Law, see Peripatetic law, Fall of bodies, etc.
Lectures, 12-13, " extraordinary ", 14, " ordinary ", 15, 17-19
Leonardo Pisano, see Fibonacci, Leonardo
Libri, G., 29
Light, 51
Lineriis, Johannes de, see Johannes de Lineriis

Lo Vasco, A., 25
Locatellus, Bonetus, 31 *et passim*
Logic and logical, 12, 102
Lombard, Peter, *Sentences* of, 172
London, British Museum, 30, 151
Lopez, P. A., 40

Macray, W. D., 149
Madan, F., 174
Mahavira, 145, 156, 158, 163, 165
Maimonides, 36
Maiolus, Laurentius, 36
Man, 64
Mariani, M., 18
Marliani, Alberto, 11
Marliani, Daniele, 11
Marliani, Gianfrancesco, 11
Marliani, Giovanni, Augustinian
   monk, 14
Marliani, Giovanni, physician to the
   Sforza, works, Collected writings,
   22, 24, 28, 30, 32-33, 139, *De anti-*
   peristasi, 31, 79, 91, 92-100, *De*
   caliditate, 14, 21, 31, 33, 79-92, *De*
   febribus, 31, *De instanti*, 27, 33,
   106, *De minutiis (et Algebra)*, 29,
   33, 145, 151-167, *De motu locali*,
   26-28, 32, 102-24, *De proportione*,
   17, 20, 26-30, 32, 125-44, 162, *De*
   reactione, 22-24, 32, 34-58, 138, 168,
   Difficultates, 28, 32, 138-39, *Dis-*
   putatio de reductione, 25, 32, 59-
   78, 97, 168, *Expositions on Avi-*
   cenna, 31, *In defensionem*, 24-25,
   32, 55-57, *Liber conclusionum di-*
   versarum, 25-27, 28, 89, 138, *Trac-*
   tatus physici, 25-27, 106, 169
Marliani, Giovanni, physician to the
   Visconti, 31
Marliani, Luigi, 11
Marliani, Paolo, 12-13
Marliani, Pietro Antonio, 12-13
Marliani, Pietro Anonio the
   younger, 13
Marliani, Pietro Camillo, 13
Marliani, Raimondo, 12
Marre, A., 146
Marsilius of Inghen, 40, 42-48, 51, 54,
   60-61, 136, 139
Marsilius of Sancta Sophia, 81
Mass, 70, 72, 78, 82, 125, 168
Mastix, 35
Mathematics and mathematical, 12,
   20, 77, 125, 139, 145-67, 168-70
Matter, 55, 61
Maximilian, German emperor, 11
Mechanics, 125 ff., 169
Medicine, 17-21, 33, 65, 83
Medium, 42, 51, 73-74, 100, 127-28

Members, human, 81, 83, 88, 90-92,
   97
Memorie e documenti (of the Uni-
   versity of Pavia), 12 ff., 31
Messinus, 140-41
Metabolic rate, 79, 83, 85
Meyer, K., 34, 93
Michalski, K., 126, 172-73
Middle Ages, 34, 36, 93, 101, 138
Milan, Biblioteca Ambrosiana at,
   29, College of Physicians of, 11,
   14, Franciscan library at, 25, Uni-
   versity of, 16-17
Mirror, 51-52
Momentum, 125
Moody, E. A., 7, 130-31
Morigia, Paolo, 11 ff.
Motion, of alteration, 36, 45, 58,
   local, Caps. V-VI, 20, 45, 50, 58,
   99, natural and violent distin-
   guished, 126-27
Munich, ms. on fractions at, 147, 152,
   155, 157, 161-62, 165

Nagl, A., 149
Nardroff, Robert von, 7
Nemorarius, Jordanus, 131, 133, 139,
   147-48, 152-57, 162, 165, 167
Newton, Isaac, 125-26
Nicoletti, Paolo, see Paul of Venice
Numerator, 151 ff.

Odo, the abbacist, 153
Oenopides of Chios, 93
Oil, 76-77
Oresme, Nicolas, 7, 102, 105, 111-12,
   120-21, 133-36, 139, 169
Organs, 88, 92
Oxford, University and schoolmen
   of, 25, 40, 101-02, 131, 134, 172,
   174-75, Bodleian library, 149, 151,
   174-75, Magdalen College library,
   174, Merton College, 174-76, Oriel
   College library, 172-73

Pacioli, Luca, 150, 154-55, 160, 162
Padua, University and schoolmen of,
   40-41, 65, 80
Paris, Bibliothèque de l'Institut at,
   144, Bibliothèque Nationale, 29,
   131, 144, 153, 171, 173, 175, schoolmen of, 38, 40, 50, 101-02
Parodius, Jacobus, 13, 31
Paul of Venice, 41, 47, 59-64, 130,
   135-36, 140
Pavia, University of, 12-22, 30-31,
   41, 65, 66, 80, 81, College of
   Artists and Doctors, 18, Collegio
   Marliani, 12, library of, 22, 24-27,
   106

Pelacani, Biagio, see Blasius de Parma
Pendulum, 140
Pepper 67, 69, 73
Peripatetic law of motion, 36 ff., 45, 56, 58, 66, 125-44, 169
Peter of Abano, 35, 80
Peter of Mantua, 27, 33, 106
Peurbach, Georg von, 148, 150, 154-55, 160, 167
Philip I, king of Spain, 11
Philip II, king of Spain, 13
Piloponus, Johannes, 125
Philosophy, 12 ff., 20-21, 33, moral, 13, 27, natural, 13, 14, 17, 27
Physics, 14, 17, *et passim*
Physiology, 49, 79, 82, 83
Piacenza, University of, 80
Picinelli, Filippo, 31
Place, natural, 61, 97-98, 126-27
Planes, inclined, 140
Plimpton library, New York, 149
Policletus ex Ferrariis, 65-66
Politus, Bass(i)anus, 40-42, 104
Pomponazzi, Pietro, 39, 42-43
Pores, 44, 46, 84, 87, 92
Potentiality, 61-62
Powers, active and resistive, 36-39, 41 ff., 46-48, 51, 54, 60, 66, 70, 74, 88-89, 98, 125-44, 169
Prime numbers defined, 159
Progression, geometric, 132, 136
Projectiles, motion of, 126-27
Proportions, 36-39, 41 ff., 66, 109 ff., 113, composition of 133-34, 136-37, 169, of motions in velocity, 125-44, mathematical theory of, 131-34, 139-41
Ptolemy, 21

Qualities, 34-38, 41 ff., 46, 54-59, 61, 64-65, 79, 81, 89, 92, 94-100, 129, 169
Quartan fever, 22

Rain, 93-94
Ramanujachariar, N., 145
Rate, of change of velocity, 119, of heat actions, 38-39, metabolic, 79, 83, 85, and see Speed, Velocity, and Acceleration
Rath, E., 150, 155, 162
Ravaisson-Mollien, Charles, 144
Reaction, 23-26, 34-58, 95, 138, 168
Reduction of hot water, 25, 32, 36, 59-78, 97, 168
Reflexi, rays, 95
Reflexion, of heat and light, 51-53, 95-96, 99-100

Reguardati, Benedetto, 20
Republica Ambrosiana, 15, 17
Rest, 111, 117, 128, 132, *et passim*
Ricardus de Versellys, 101
Richard of England, 148, 151
Rigveda, 151
Rome, Vatican library, 30, 172-74, 176, Vittorio Emanuele library, 173
Root, extraction of square, 166-67
Rosetti, Roger, see Royseth, Roger
Royseth, Roger, 172-73

Sancto Nazario, Jacobus de, 32
Santa Maria delle Grazie, 12, 21
Savonarola, Michael, 66
Sbaraglia, G., 40, 172-73
Scholasticism and schoolmen, 38, 40-41, 59, 72, 78, 101-02, 124, 128, 130, 139, 141, 167, 168
Scotus, Octavianus, 31 *et passim*
Sermoneta, Johannes, 79-80, 84-85, 88, 92, 169
Sex inconvenientibus, see *Auctor de*
Sforza, Francesco I, 15, 17, 18, Francesco II, 13, Galeazzo Maria, 21, 31, 99, Gian-Galeazzo Maria, 20, Lodovico il Moro, 11, 13, Massimiliano, 11
Siena, University of, 41
Silberberg, M., 147
Similars, 67-68, 70, 77-78
Simples, medicinal, 34-36
Singh, A. N., see Datta, B., and
Sitonis, Johannes de, 14, 29
Skin, 87
Smith, D. E., 150
Smoke, 98
Soto, Dominic, 101
Soul, 82
Space, a moving body defined by, 103-104, 104-24, 169
Speed, law of uniformly varying, 101-24
Spiritus, 82, 84-87, 90-92
Sridhara, 145, 156, 165
Stagirite, see Aristotle
Steam, 98
Steinschneider, M., 148
Stifel, Michael, 165
Stomach, 81
Subterranean places, 97
Sudhoff, K., 35
Sun, 51, 94
Superior influences, 61
Supernatural, 100
Surface, concave, of water, 98
Suter, H., 146, 154
Swineshead, John, 175-76

Swineshead, monastery of, 175-76
Swineshead, Raymond, 173
Swineshead, Richard, 7, 23, 24, 25-27, 32, 40-41, 44-46, 49-50, 53, 70, 102-03, 105, 111-12, 116-17, 119, 124, 136, 140, 144, 169, 173-76
Swineshead, Roger, 103-04, 171-76
Swineshead, William, 171

Taylor, N. B., 83
Tealdo, Lazaro, 20, 81, 99
Temperamentum, 90
Temperature, 34-35, 58, 59, 63, 65, 68-69, 72, 74, 76, 78-79, 82-85, 93, 95-96, 168, body, 72-92, 169, and see Degree, Heat and cold, intensity of
Thermometer, 58, 59
Thomas Aquinas, 34, 130
Thomas de Garbo, 36
Thorndike, L., 7, 170, *Magic and Science,* 25, 36, 65, 82, 131, 149, 173, 175, *Catalogue of Incipits,* 30, 36, 151, *Science and Thought,* 150, 170
Time, in measuring velocity, 101, 103 ff., 124, 128
Tiraboschi, G., 21
Tissues, 83
Torni, Bernardo, 19
Touch, organ of, 100
Tropfke, J., 146, 154, 165

Ueberweg, F., 40
Ugo de Benziis, see Hugh of Siena
Umasvati, 160
Uselis, Richard, see Ricardus de Versellys

Vacuum, 64
Valentinelli, G., 24, 28, 174

Valla, George, 22, 81
Vapors, 46, 64, 68, 71-72, 75-76, 99
Vatican library, see Rome
Velocity, kinematic study of, 101-24, dynamic study of, 125-44, 169
Venice, San Marco library, 22 ff., 174, *et passim*
Vergellis, Richard, see Ricardus de Versellys
Victoria, 130
Vienna, University of, 40
Viglevani (Vigevano?), 99
Vimercati, Giovanni, 16
Vinci, Leonardo da, 22, 144, and see Duhem, P., *Études sur Léonard de Vinci*
Visconti, 11, 15, Filippo Maria, 13, 15, Galeazzo, 31, Galeazzo II, 13, Gian-Galeazzo, 11, 15, 31
Vittori, Benedetto, 144
Volta, Zanino, 11 ff.
Volume, 70, 82

Walter of Burley, 38, 40-42, 49, 60, 62-64, 78, 130, 136
Wappler, E., 147, 149, 166
Water, heating and cooling of, 43, 46-49, 55, 59-78, 93-100, 168
Weight, 127-28
Weileitner, H., 120-21
Widman, Johannes, 150, 155
Woepcke, F., 146-47
Wohlwill, E., 125-26
Wycliff, John, 140-41
Wylliott, John, 174

Zarkali, 148
Zarotus, Antonius, 31, *et passim*
Zonta, C., 80
Zorzanello, P., 22, 24